THE TRINITY CIRCLE

SCIENCE AND CULTURE IN THE NINETEENTH CENTURY

Bernard Lightman, Editor

THE TRINITY CIRCLE

ANXIETY, INTELLIGENCE, *and* KNOWLEDGE CREATION *in* NINETEENTH-CENTURY ENGLAND

WILLIAM J. ASHWORTH

University of Pittsburgh Press

Published by the University of Pittsburgh Press, Pittsburgh, Pa., 15260

Manufactured in the United States of America
Printed on acid-free paper
10 9 8 7 6 5 4 3 2 1

Cataloging-in-Publication data is available from the Library of Congress

ISBN 13: 978-0-8229-4687-8
ISBN 10: 0-8229-4687-4

Cover art: Stipple engraving of William Whewell, courtesy of Wellcome Library, no. 9673i; detail of *Michelangelo's Creation of Adam.*
Cover design: Alex Wolfe

CONTENTS

Acknowledgments vii

PROLOGUE 3

INTRODUCTION 6

CHAPTER ONE. Trinity in the Early Nineteenth Century 18

CHAPTER TWO. Putting the Soul Back in the Machine 51

CHAPTER THREE. German Biblical Criticism and Guessing 64

CHAPTER FOUR. Trinity and Religious Dissenters 82

CHAPTER FIVE. An Ethical Economy 94

CHAPTER SIX. The New Master of Trinity 101

CHAPTER SEVEN. Genius and Great Scientific Discoveries 109

CHAPTER EIGHT. Trinity and Moral Philosophy 121

CHAPTER NINE. Space, Time, and the Idea of a Cambridge Education 130

CONCLUSION. The Last Days of the Trinity Circle 151

EPILOGUE. The Demise of God and the Rise of Machine Intelligence 164

Notes 173

Bibliography 225

Index 249

ACKNOWLEDGMENTS

I was never comfortable with the core subjects taught at school. They always seemed rather artificial and contrived. This in turn led to my interest in the sociology of knowledge and ethnomethodology. Here I was fortunate to have the guidance of an excellent and thoughtful teacher, Graham Button. I first became interested in the history of knowledge under the guidance of the brilliant historian of science Crosbie Smith, who first alerted me to the emerging and changing nature of natural philosophy during the nineteenth century, particularly through the work of John Herschel. I then worked for eighteen months in the City of London as a dairy trader, where I quickly learned that the only knowledge that mattered was anything that would lead to a profit. My only success was getting concentrated butter into Tesco.

I had to change tracks. It was time to go back and try to understand how we had gotten to such a worldview—at least in England. I subsequently embarked on a PhD under the supervision of Simon Schaffer. It was here that I documented the breakdown of natural philosophy into various specialized scientific disciplines. My focus was, to a certain degree, on Herschel and Charles Babbage. It was in their work, and like-minded folk, that one could see the rise of an industrial approach to knowledge creation during the nineteenth century, and a shift in faith from God to humans and, ultimately, machine intelligence.

What I did not truly appreciate was the backlash among predominantly High Church Anglicans toward this collapse in traditional epistemology and social authority. It was only during the final third period of my PhD, when employed by Trinity to map the extensive William Whewell archive, that I realized there was a robust alternative perspective. Whewell and his close Trinity circle did not fit the usual depictions of intelligence and the discovery of knowledge portrayed in the main historiography. Luckily I was able to attend lectures given by the foremost historian of Christianity and politics in Britain during this period,

Boyd Hilton. We have subsequently become friends, and he continues to offer creative comments that have greatly aided this book.

From these origins I departed the subject to concentrate upon British industrial history and the role of the state. This culminated in two books on the topic. I then, quite suddenly, returned to the subject while commencing a chapter on Trinity College during the nineteenth century. The materials—both primary and secondary—were so rich that the work soon mutated into this book. I have encountered numerous people along the way who have been vital in helping to shape this work.

More than anyone else it was Simon who contributed and inspired me the most in trying to understand the history of human knowledge. His omniscience and wit are too well known to underline here. He has been a wonderful supervisor and good friend. A special thanks must also go to Crosbie, Boyd, Rob Iliffe, Jim Moore, Iwan Morus, and Andy Warwick for their generosity and encouragement over the years. I have been lucky to have had feedback from others, in particular Jim Bennett, Bob Brain, Dan Brown, Alexandrina Buchanan, Jenny Bulstrode, Brian Dolan, Celia Donert, Menachem Fisch, Graeme Gooday, Arne Hessenbruch, the late Jeff Hughes, Michael Hughes, Myles Jackson, Adrian Johns, Stephen Kenny, Kevin Knox, the late Kate Marsh, Richard Noakes, Jim Secord, Keith Spendiff, Larry Stewart, Christopher Stray, and last but not least, the late Alison Winter. Throughout this time a continual source of stimulation has been my friendship with Erica Charters, Len "Boss" Rosenband, and Andre "Flipflop" Wakefield. Nothing will replace that year at the Dibner Institute—then located at MIT in Boston (2001–2002).

I could thank numerous institutions and their staff but one stands out by far. None of the following would have been possible without the people and archive at the Wren Library, Trinity. The archivist, Jonathan Smith, has been a constant help throughout the writing of this book. I was also lucky enough to be awarded a short-term fellowship at Trinity that enabled me to complete my study of Whewell's sermons. I am indebted to the feedback from the editor of the Science and Culture in the Nineteenth Century series at Pittsburgh, Bernie Lightman, the historians who sit on the editorial board, and the anonymous reviewers. A special thanks must also go to the professionalism and efficiency of Abby Collier, the manuscript editor for this volume at the University of Pittsburgh Press, and the copy editor Judy Loeven. The process could not have been smoother. Finally, my family has been vital. My two sons, Oliver and Harvey, have kept me both grounded and entertained. Throughout my sister, Kerrie Spendiff, has been there for me. I am deeply grateful to her frequent wisdom and unconditional love.

THE TRINITY CIRCLE

PROLOGUE

Today in large swaths of the West, we live in a much more secular world that typically denies the rule of a Christian religion and certainly an active intervening deity. Faith is now far more invested in human reason and, more recently, machine intelligence. This is historically a new phenomenon.[1] Nonetheless the debate over life, nature, and epistemology continues to rage. This is certainly the case in parts of the United States where the argument over creationism and its evangelical biblical roots remain dominant, especially in regions of the South. The idea that everything randomly evolves through natural selection and without divine intervention is, for these people, symptomatic of our Fall and corruption.[2] The consequence has been the eclipse of the latter and a complete trust in our species's reason. Faith in God's providence and design has been replaced by us. The result, for these Christian evangelists, is an immoral and materialist world devoid of the true Creator.[3] This has not only created a false knowledge, they argue, but an accompanying immoral world designed by an artificial human-corrupted intelligence—from education and commerce to our lifestyle, knowledge, and politics. There is still a great deal of sympathy for such a view. A Gallup poll in 2017 claimed 38 percent of the US population believed humans were created in their current form by God sometime in the last ten thousand years; the percentage might be sinking but the arguments remain extremely pertinent.[4]

One major aspect to this debate has been the alleged demarcation between empirically measured fact and some vague divine metaphysical morality. A recent, now deceased, writer tried to bridge the worlds of science and religion by arguing, in the end unsuccessfully, that the former deals with material reality and the latter with morality, what he termed NOMA (Non-Overlapping Magisteria).[5] As we shall see in the following history, this was a severance that haunted certain English nineteenth-century commentators. If morality was not divinely within us, it was simply shaped by our changing material needs. For the mainly social elite defenders of the former you could not have true knowledge without true morality; both derived from established Christian faith and an active intervening God.[6]

To grasp true knowledge required a transcendental symbiosis between the inner human (spiritual) and outer physical (material) domains; there were certain eternal and divine truths (ideas) in both the material and moral worlds. The guide was Scripture, the word of God, as taught through the Church of England. By contrast, a view was growing that emphasized the greatest happiness, reducing everything to numbers. This came to be seen as neutral in decision making, therefore making policy justifiable: those who controlled the numbers controlled the power. This movement has broadly been associated with the rise of what came to be termed *utilitarianism*, a term popularized by the political philosopher John Stuart Mill. In many ways this has become the bedrock of modern Anglo-American society.[7]

The real question was what kind of intelligence and knowledge was this social movement trying to produce? This is intrinsic to the subsequent prescribed social and material environment. The momentum, as we shall see, was less an intellectual debate and more one embedded in traditional social relations and a clamor for social reform. For human reason to be true and let loose, an increasing number of campaigners argued, we needed to be emancipated from the debilitating grip of religion (meaning primarily the Established Church and its relationship to the state) and the creed's accompanying final causes; nonmaterial explanations were to be crushed. The passions were condemned, and the pursuit of material gain became a moral attribute; decision making, duty, and morality were to be based on self-interest and rational calculation. However, this inversion of traditional customs was not an obvious outcome during the nineteenth century and was by no means certain. Indeed, its legacy still dominates the shaping of our notion of cognition and morality.[8]

This book's focus is on the demise of the traditional teachings of the Church of England and all the varying attempts to defend this institution. As we shall see, for those at the center of this book, Scripture, morality, and knowledge were all entwined. Nor did they like the traditional Anglican emphasis on natural theology or, if you prefer, intelligent design in the William Paley sense. This was now seen by many as complementary to the rise of a mechanistic utilitarian worldview.

Indeed, the traditional Anglican emphasis on intelligent design was as antithetical to the people at the heart of this book as the secular material outlook that took hold. After all, as conservative Anglican Tories increasingly moaned, the two perspectives complemented each other. Rather, discovery and access to divine objectives were a combination of observation and the imagination, a spiritual-metaphysical-inductive method rather than solely an empirical and/or logical deductive process.[9] Recent historians have attempted to locate the origins of modern liberalism in Britain. They wrongly claim that this was due to the country's embracing of freedom and creativity, unlike the authoritarian European Continent. For Otto Mayer this led to the death of the clock/machine metaphor in nineteenth-century Britain. He concludes, "The neglected qualities, private enterprise and individual initiative, are of course, precisely the qualities that liberal political systems needed to cultivate."[10]

On the contrary, it was in nineteenth-century Britain where the machine metaphor—if not more than a metaphor—became arguably further embedded than anywhere else in the world. This is a history of one small corner of the globe where all this was fiercely fought. It centers on, but not completely, Cambridge University and especially Trinity. Rather than an age of improvement, it was for these people far more an age of anxiety.

INTRODUCTION

> There exists a Supreme Mind, which is the cause and foundation of the universe. Those who hold this, and who also hold that the human mind can become possessed of necessary truths, if they are asked how it is that these necessary truths are universally verified in the material world, will reply, that it is so because the Supreme Creative-Mind has made it so to be:—that the truths which exist or can be generated in man's mind agree with the laws of the universe, because He who has made and sustains man and the universe has caused them to agree:—that our ideas correspond to the Facts of the world, and the Facts to our Ideas, because our Ideas are given by the same Power which made the world, and given so that these can and must agree with the world so made.
>
> William Whewell, 1860

The subject of this book is a quest, by a group centered on Trinity College, Cambridge University, to ground a human purity informed by divine faith directed by the Established Church. This was, certainly at a socially elite level, fundamental to objectivity and came from within and not from outside. It was also a fierce critique of the rise in utilitarianism and individualism, in short a now largely forgotten theological condemnation of values associated with a rapidly colonizing capitalism in nineteenth-century England.

The words in the opening epigraph were written by William Whewell, son of a Lancaster house carpenter and a mother who annually contributed enigmas and charades to the *Lady's Diary*.[1] The argument was the mature articulation of several decades of intellectual reflection behind the walls of his beloved alma mater.[2] In a sense, it was a traditional English variant on a prevailing German idealist argument, which emphasized that true understanding was only possible because certain ideas explaining the universe were somehow within us. To be sure this was very much an English hybrid version in which natural philosophy was subsumed to religion; access to underlying ideas of the world could be sourced only via true religious faith—more Lutheran than Kantian—guided by a state-supported National Church. For Whewell, such truth was a divine sensory-metaphysical synthesis.[3] An "Idea" represented "those inevitable general relations which are imposed upon our perceptions by acts of the mind, and which are different from anything which our senses directly offer to us." Examples included "space and time, number and figure, cause and resemblance," which Whewell termed "*Fundamental Ideas.*"[4]

The place where the "Supreme Mind" could be mirrored was the apex of Anglican education, Cambridge, and especially Trinity. Any notion of a recognizably modern science was still at least nearly a century off. Religion still infused, directly or indirectly, all the ways of elite knowing. Even those that denied the relevance of religion had to labor extremely arduously to do so.

The rise of secular faith, political radicalism, science, commerce, and industry was, as it seemed to Anglican critics, undermining this spiritual world and challenging it with a superficial material one: a human-centric rationalist society hell-bent on measurable betterment via profit, consumption, and a prevalent notion of progress. There was no room in this movement for the soul. This was an entity distinct from your body—the seat of your mind, character, thoughts, and feelings. Without a soul, you were incapable of redemption from the power of sin through divine grace. Salvation was fundamental to interpreting the world. The absence of the soul, for people at the center of this book, created a false dialectical relationship between the human inner domain and the outside creation; the result was a cold mechanical world devoid of a divine spirit. For others it was simply a sign of ignorance. The harbingers of this new, purely materialistic world wanted the social system and political constitution to be reformed, with the old certainties such as the state and Church of England coupled with the traditional sinews of political power reformed. For Whewell, the potential collapse of the prevailing creation was a chilling vision he shared with several close friends at Trinity and beyond.

This book tells the history of a struggle, spearheaded from Cambridge by a Trinity Circle (TC), between faith in God versus trust in pure human reason (a priori deduction) and/or human sensory empiricism (experience)—both of the latter two approaches, for the TC, were devoid of God; faith purely in human sensory observation and/or abstract human-centric rational intelligence sent you down the path to perdition, a false world via an illusory logic. Authority came from above and not below. An emphasis on certain innate ideas was informed by the philosophy of Immanuel Kant; reason could not work without ideas (for example space and time), and ideas were not, contrary to John Locke and David Hume, simply derived through the senses. Divinely planted ideas, for the TC, were the foundations to true knowledge. This was the attraction to Kant's transcendental idealism and emphasis on Euclidian geometry; our understanding of the nature of the world was not possible without both innate ideas and experience. Central to this Kantian belief was an emphasis on a morality informed by beauty, feelings, and the imagination. This, for Kant, was the recipe for genius.[5] The TC, of course, shared overlaps with peoples and groups beyond the college's boundaries. However, they were a distinct and significant intellectual circle in defining an important English nineteenth-century view toward a notion of knowledge.

The term *Trinity Circle* (TC)—rather than, say, *ideology* or *philosophy*—is deliberately coined for the following reasons. Despite the varying intellectual tools adopted by the leading protagonists of this book, they all sought to defend and strengthen the Church of England. All agreed that the base of knowledge was God and that the only way to preserve this divine centrality to life was via a robust Established Church and accompanying Anglican education. As such they all aggressively disliked the new abstract deductive (Ricardian) political economy, growing materialism, Benthamite utilitarianism, abstract French algebraic mathematics, evolution, and a growing faith in human logic. In short, they sought to crush anything that challenged traditional social authority and the Church of England. They all had, to begin with, overlapping and different approaches to how best to achieve this objective but were all united in their attempt to do so. This is hardly surprising at the commencement of the nineteenth century since tensions were high after the French Revolution and the disruption of the rise of an industrial society. Questions concerning the nature of knowledge and traditional authority were in a state of flux.

A unified Established Church was the solution to the socioeconomic instability of the post–Napoleonic War crisis. The commercial spirit, political radicalism, and a swath of blasphemous and seditious publications needed to be stopped. The traditional alliance between the Church, Constitution, and landed classes needed to be bolstered. This alliance was the true guardian of cultural authority and desperately needed to reassert itself. The TC would draw on anything they thought would further this quest. The organic unity of society was being destroyed by a false empiricism and deductive logic that had no basis. The question of how best to do this haunted numerous Anglicans, but there was no consensus in the Church of England on how to achieve it. This was also reflected in the approaches taken by the TC. Nonetheless, all were united in the belief that truth came only via knowing God from within. They were determined to defend the prevailing cultural authority and underline the divine origin of the human species.

The journey for Whewell began in October 1812 when he arrived in Cambridge to commence his university education at the bottom of the social scale for entrants, namely a subsizar (literally a subsidized student). At the top of the social scale were noblemen, who were admitted due to their background rather than academic ability.[6] Whewell seemingly overcame any worries concerning his social roots and forged a close friendship with another Trinity student, Julius Charles Hare. By contrast Hare came from a very wealthy aristocratic family; he was the son of the famous historian Francis Hare-Naylor and the painter and cousin of the Duchess of Devonshire Georgina Shipley. Hare was born in Vicenza, Italy, before being educated at Charterhouse School (then based in Smithfield, London), and came to Cambridge with the more typical, privileged status of a middle-ranking pensioner who paid full tuition fees.[7] It was while in Italy that he was greatly informed by Clotilde Tambroni, an Italian philologist, linguist, and poet. Between 1793 and 1798 she was a professor of Greek language at the University of Bologna, and from 1800 a professor in Greek and literature.[8]

The following year Whewell and Hare bonded with Hugh James Rose who also entered Trinity as a pensioner. He was educated at Uckfield School in Sussex where his father was headmaster. Already by the age of four he could recite the Greek alphabet after private lessons from the former Uckfield pupil and Jesus College, Cambridge, student Edward Daniel Clarke.[9] They were joined a year later by Connop Thirlwall, son of a clergyman, who came from East London, again as a pensioner. Thirlwall already knew Hare from his time at Charterhouse, and it was not long before they were reacquainted. The school, particularly under the auspices of its headmaster, Matthew Raine, a former Trinity fellow, had become a prominent feeder of boys to his former college. Whewell, Hare, Rose, and Thirlwall went on to forge a distinct TC. To be sure there were levels of disagreement but enough shared family resemblances to shape a coherent and important group. Their glue, as opposed to approach, bonded them to safeguard the Church and not break away from tradition in the sense of the later Oxford Balliol idealists. The emphasis of the TC was on the next world and not this one.[10] Last, many of their views were shared by a slightly older Trinity fellow, the Yorkshire son of an Anglican vicar, Adam Sedgwick, who was educated at Sedbergh School in Cumbria—the same seminary where the Romantic poet William Wordsworth sent his sons. We shall see that the sinews of this core circle, and indeed other Trinity graduates, stretched far and wide.[11]

All the TC came to be united in their quest to demonstrate that "mankind," with an accompanying evangelical emphasis on human reason, was secondary and always dependent on an all-powerful and actively interventionist God. This had clear implications for the nature of religious faith, knowledge, and traditional social authority. Access to genuine understanding, they believed, was only possible to those with a soul; this could not be easily articulated but was crucial to comprehending God's world—be it knowledge, social structure, the economy, or, crucially, morality.

The TC were all united in their defense of the Church of England. This manifested in a shared view toward many issues they believed were highly significant, particularly concerning the future of the Church, knowledge, and the status quo. However, in terms of how best to preserve the Church they often greatly differed. At one end of the spectrum was Rose and the other Thirlwall. The former fervently disliked the early enthusiasm by the latter for German scriptural criticism and toleration of religious Dissenters. Here Rose was far more representative of the prevailing Anglican High Church status quo than Thirlwall. Since the French Revolution there had been a reaction against anything foreign. German biblical scholarship was now associated with a threat to the truth and validity of the Bible—thence Christianity and the Church of England. Since the late 1790s journals like the *Anti-Jacobin Review* slammed German scholarship as antireligious and a threat to the divine inspiration of the Bible. It was important to underline stability and at the heart of this was the Bible. In 1804 the British and Foreign Bible Society loudly asserted it only recognized the King James Bible as "the translation of the Scripture established by Public Authority," that is, the Church of England.[12]

By the 1850s Thirlwall would be much closer to Rose's tough and uncompromising stance toward the divine basis of Scripture. The use of such German criticism came at a particular instant in their debates, but by the publication of *Essays and Reviews* in 1860, it had become an adverse source. Ironically the young Thirlwall and Hare's promotion of German biblical scholarship would by this time be seen, as Rose predicted, as deeply erosive to the Church and Christian religion. This was typical of the TC, all of whom tended to abandon any resource their enemies found helpful—from German criticism to, as we shall see, mathematical algebraic analysis and human logic.

The Trinity Circle engages with the TC's concerns and how they attempted to promote a historicist version of the Anglican faith. This in turn had to somehow be embodied within the academic curriculum at Trinity and then Cambridge. One general feature of this was working out how to put the right moral philosophy, which mirrored this quest, into such a syllabus. For spatial and archival reasons Whewell represents the spine of this history. First, he was the only one of the TC who spent his whole life at Trinity and, indeed, literally became the outward embodiment of the college as master in 1841. Moreover, his records, letters, and output have been by far the best preserved. In this sense Whewell is the hub and the rest are spokes within the following history. This is not to detract from the others, but for these reasons Whewell was at the geographical and intellectual heart of the TC. Nonetheless, this is not a biography of Whewell or indeed the TC. Rather, it is a moment that illuminates a controversy over both the nature of religion, scientific belief, and an emerging capitalism. Far from being an age of improvement, it was an age of anxiety for these men.

To be sure, the TC was not always a perfect circle and their views clearly differed in how to achieve a National Church. Nonetheless their differences tended, wherever possible, to remain private. On the occasion they did go public, such as Thirlwall's vocal support for Dissenters to be allowed into Cambridge, a solution of sorts would be found. Rose stayed resolute in his doctrinal beliefs all his life. From the 1820s Whewell also became far more static and defined in his outlook. Hare, Thirlwall, and Sedgwick were perhaps the most liberal during this decade. However, after the Great Reform Act of 1832 this started to change, with Hare taking up his family clerical living at Herstmonceux, Sussex. Here he became far more anti-change and suspicious, if not fearful, concerning the future of the Church, the Anti-Corn Law Movement, and Chartism. Thirlwall and Sedgwick continued to support the move to allow Dissenters to take a degree at Cambridge and Oxford. However, by the 1850s they had also become much more conservative and reactionary toward ecclesiastical and social reform. By this point Thirlwall had assumed one of the top positions in the Church when he was appointed bishop of St. David's.

As well as seeking to preserve a National Church, by the 1820s all the TC were suspicious of human-centric rational reformers, especially revolutionary French philosophers, abstract analysts, and British utilitarians. These people were selfish individualists and dangerous radicals committed to tearing up Albion's organic structure and severing the sacred relationship of the Church and state. Established religion should always have the upper hand, correctly taught via an accompanying doctrine, over all other forms of belief. The National Church was the standard of measurement (knowledge, morality, and authority) that secured social order from political chaos and insubordinate challenge; moreover, it was the key vector in ensuring the safe passage from this temporary world to the eternal future in the next. This theology, and an accompanying hierarchical social and moral structure, needed credibility and obedience. However, the means to protect these essential axioms would come to divide nineteenth-century England and, for a time, the TC. There is no doubt that it was during the volatile and divided 1830s that much of all this came to a head; it is no exaggeration that the future of the social hierarchy was at stake.

The traditional Anglican relationship with the state was under mounting threat, and as such, the nation's Constitution and the Established Church needed to be revitalized by spreading an authentic religious temperament. This, it was hoped by many fearful of radical reform, would ensure that the prevailing quest for change—built on false and godless principles—would become redundant. Severing the dominant role of the Church and its educational seminaries, Oxford and Cambridge, had to be rigorously guarded against. The prevailing "steam intellect" that was dangerously heating up the country should not be allowed to enter Anglican seats of learning.[13] Instead the emphasis had to be on finding one's soul—preparing for the next world—and not materialistic ambition in this one; this was the bond of Church unity and security for the country's ruling elite.

Whewell later confessed to his brother-in-law, James Marshall, his love for "our National Constitution and in our National Religion. I believe that these embody more of the truth, are better approaches to the true form of Church and State, than have ever yet been established." The National Church and Constitution were "living things" and not "mere formulae"; the implicit relationship between the Church of England and the Constitution was not negotiable. Reform should, and could, take place only within Anglican terms.[14] Along with the contrasting Oxford University theological movements such as the Oriel Noetics and later the Tractarians, the work of the TC was popular with a vast respectable swath concerned with theology, epistemology, morality, and social authority.[15] There was much overlap but also huge differences in such perspectives, in particular "the Noetics' logical speculative approach to political economy."[16] Although the two groups shared the same objective of preserving the Church, their approaches were antithetical to each other.[17]

Chapter 1 looks at Cambridge and the environment the TC was entering in the early nineteenth century. This was an exciting and turbulent time, and proof of the vulnerability of the Church of England and God's rule was not difficult to see. For the first half of its first decade the British were still waging war with the godless Napoleonic regime. France's traditional establishment had been stripped away and replaced by an ahistorical and, for many frightened English people, false human-centric rational structure. The French Revolution (1789–1799) had released a hurricane of change with competing forms of knowledge challenging traditional theological teachings. For Anglicans in England the indigenous equivalent meant—most threateningly—utilitarianism, die-hard empiricists, the importation of French algebraic mathematics, a priori logicians, religious dissenters, and a radicalized working class; it spelled the possible end of the Established Church, morality, and traditional authority.[18]

Ecclesiastical reform had been debated for several years, and its importance was magnified during the French Revolutionary (1792–1799) and Napoleonic Wars (1803–1815). The privileges and monopoly of the Church had given rise to a large wave of critical condemnation. To add to the woes, a strong Christian evangelical movement had also taken root, insisting that the Church desperately needed to reform. For example, they insisted that clergymen who did not stay in residence where they worked should be prosecuted—this was subsequently legislated in 1803. Moreover, they challenged the need for the Church's doctrines and practices to be codified in the 39 Articles.[19]

The Anglican Church's clergy saw their role as thwarting the rise of French radicalism and religious nonconformity. However, they themselves had become a source of contempt for many residing in the country. For example, the tithe—a Church tax—had especially become a source of rural dislike. This led to another parliamentary bill in 1808 to improve the maintenance of the Church's clergy. The situation was seemingly irresolvable, since resources to provide an adequate stipend for a curate to live in an appropriate dwelling and have a married life were already hard. Hatred of the tithe led, two decades later, to the clergy being attacked during the Swing Riots.[20]

Threats were also coming from elsewhere. Cheap and accessible publications drawing on astronomy and new disciplines like chemistry, biology, and geology were harnessed to aid the march of political reform. For example, aspects of the work of the French naturalist Jean-Baptiste Lamarck were harnessed by the London medical and penny presses to demonstrate that we had all transmuted from self-acting matter devoid of any divine intervention. This, as Adrian Desmond puts it, raised "the spectre of design without a designer."[21] Accompanying this was also the breakdown of the organization of natural philosophy from its traditional institutional home at the Royal Society of London (est. 1660) to satellite specialist groups like the Geological Society of London (est. 1807) and the Astronomical Society of London (est. 1820). A few years later the reformers gained a breakthrough in higher education with the creation of London University in 1827. Henry Brougham two years earlier had

Robert Seymour, "Satire against corruption: A huge automaton representing the new London University (from 1836 University College, London and, finally, from 1907 University College London) tramples over greedy clerics, doctors, lawyers and the crown," 1828.

predicted to the radical MP Sir Francis Burdett that such an event would crush "the High Church Bigots."[22]

Just as religious sectarianism had triggered an English Civil War in the mid-seventeenth century, for many Anglicans something similar was brewing in the early nineteenth century. Knowledge based primarily on human reason was, more than ever, challenging traditional *theological* teachings and with it the sacred alliance between Church and state.

The unified eighteenth-century trinity of politics, religion, and knowledge was under critical scrutiny like never before; the monopoly of the Church of England, with its rituals and Prayer Book as a binding force, was being challenged, leading in many ways to its revival within a deluge of social upheaval and accompanying notions of knowledge and authority. For many it reached a crisis point when the Test and Corporation Acts were repealed in 1828; the following year Dissenters and Catholics were placed on the same level as Anglicans. The 1832 Great Reform Act further expanded Parliament and threatened its essentially Anglican makeup. If the National Church was eclipsed it would open the path to diverse religious sectarianism, the tyranny of democracy, moral decay, the rise of the working classes, and possible revolution. Democracy then, as opposed to now, was for most in social authority a derisory concept. Samuel Taylor Coleridge was fearful: "The Church is the last relic of our nationality," and its fate dominated his thoughts.[23] Chapter 2 deals with these themes and the attempt by the TC to thwart materialism and retain the soul within the university's teaching curriculum; human feelings ran much deeper than mere human reason and sensory perception. Ultimately epistemology was divine and could be revealed only by someone with the right religious demeanor.

Boundaries were being remade at every level—from society, the workplace, and politics to education and knowledge. The TC was one elite group caught in this growing environment of uncertainty and change. In 1810 Thirlwall described French poetry as "tedious and insipid" and condemned home-produced radicals such as Burdett and the corrupted thinking of the Etonian, rational etymologist, and former St. John's student Horne Tooke. The French Revolution, he hissed, had simply "produced wretches hardened by crime" who convinced "themselves into the idea of a disbelief of a future state of existence."[24] All the TC were quickly dismayed at the emphasis they found placed at Cambridge on the works of Locke and Paley.[25]

Hare told Whewell in 1822: "Plato is worth ten thousand Aristotles & Locke's."[26] Tooke drew on an extreme Lockean version of language that reduced all cerebral action to human linguistic activity devoid of any divine input.[27] Locke essentially saw religion as a private matter that should be separate from a National Church and state. More precisely, Locke did not object to established religion per se but did not think it had any special legitimacy (as against the various other sects), and he did not think it was in any way identified with the state. The Church of England, of course, insisted that only via its 39 Articles could religious interpretation take place. Where the source of understanding came from within for the TC, it was external for Locke and comprehended by human sensory-based reason. This latter argument was also established by the works of prominent seventeenth- and eighteenth-century deists such as the Irish freethinker John Toland and Matthew Tindal.[28] Such deism had declined but took new forms from the second half of the eighteenth century in Socinianism, Arianism, and Unitarianism. Cambridge was still reeling from the controversy aroused after the expulsion of the King's School–educated Unitarian and onetime student of Paley, William Frend, on religious grounds in 1788 just before the French Revolution.[29] Frend was first deprived of his tutorship in 1788, but the real crisis came in 1793–1794 during the French reign of terror, regicide, and war. Within this context Frend was expelled, though he continued to derive his fellow's stipend until he married.

Paley's work, primarily his *Principles of Moral and Political Philosophy* (1785), supported utilitarian morality and placed the emphasis on man's sensory reason to detect design. It was this that was examined at Cambridge and not Paley's theology, which argued for the authenticity of the Bible, Jewish prophesy, God, and miracles. In his empirical evidence of Christianity, *Natural Theology or Evidence of the Existence and Attributes of the Deity* (1802), Paley placed a human-centric reason at the core of Christian morality and described a contented world filled with more happiness than pain. A legalistic and empirical view of theology, in which evidence of God was simply found in nature, revolted men like Coleridge and the TC.[30] As Robert Hole neatly summarizes, both Jeremy Bentham and Paley "pushed men down a secular road which replaced divinely imposed duty with selfish calculation."[31] Competitive individualism was replacing the moral duty imposed by God. This was a short step away from what Samuel Clarke had much earlier, in his public scrap with Gottfried Wilhelm Leibniz, condemned as materialism: "The notion of the world's being a great machine, going on without the interposition of God, as a clock continues to go without the assistance of a clockmaker; is the notion of materialism and fate, and tends (under pretense of making God a *supra-mindane intelligence*), to exclude providence and God's government in reality out of the world."[32] Self-interest was no basis for the organization of society and was a fiction rather than based on facts. Indeed, it was deliberately setting up a false divide between the landed interests and everyday folk.[33] The TC hated this perspective.

There were equally theological disputes erupting over the credibility of the gospels underpinning the Christian doctrine. This is the subject of Chapter 3, as well as the TC's response to this fraught situation. The key to survival for the established clergy was to resist sectarian schools of conflicting theological wrangling and become united through nurturing one's soul via the divine belief prescribed by the National Church. The traditional emphasis upon natural theology as a way of uniting different views had now, for Anglicans like the TC, become subversive and closer to materialism and a world without God. This issue would continue to dominate the lives of the TC throughout the nineteenth century. To substantiate the truth within the gospels, Hare and Thirlwall applied controversial German biblical criticism to create a credible doctrine for the National Church; a doctrine that could embrace an array of dissenting views. These two men were then at the forefront of importing German critical studies into England. Hare confessed to his mother in 1820 that he owed German literature "my ability to believe in Christianity."[34]

By this point, the TC was becoming divided. Rose, now an orthodox Tory member of the High Anglican Church, condemned the German approach as no better than any other godless rational philosophy; all the gospels came directly from the word of God and were beyond human questioning. Whewell increasingly agreed with his old strict and stiff friend. This disagreement within the TC came to a head in the refusal to allow Rational Dissenters into Trinity during the early 1830s. This event forms the subject of Chapter 4.

Thereafter, however, even Thirlwall started coming around to Whewell's Anglican conservatism and Rose's fear of German biblical studies. Ironically, just as the most fervent advocate of German biblical scholarship was turning his back on it as a threat to the Church of England and Scripture, the rest of the country was beginning to embrace it.[35] All the TC feared the growing dominance of a new political economy and utilitarian philosophy, which was emerging as a doctrine of legitimation for the materialistic world of profit and consumption. This was an abstract, human-centric, political economy that legislated axioms without any historical foundation based on inductive scrutiny. The former Caius graduate Richard Jones, hounded by Whewell, set about providing an alternative economics that he preferred to call an "ethical economy," based on historical facts spread across space and time. This ethical approach was closely tied to bolstering the Church of England—particularly its close relationship to the state and as a large landowner. Although Jones was not a product of Trinity, he was extremely close to all the TC and probably Whewell's closest friend. The attack on political economy is the theme of Chapter 5. Consequently the works of Jones and Sedgwick are also incorporated, as their labor on educational apologetics and moral economics made them important and useful to the TC but not necessarily as equal members.

The new rational knowledge was based, as far as the TC thought, on un-credible grounds and was penetrating too many institutions—including Cambridge, Oxford, the Church, and the Constitution. Someone had to make a stand and prevent such knowledge from entering the Anglican university's curriculum and the Church's teachings. Great scientific discoveries ultimately came from those in touch with innate divine ideas and manifested in the genius of pious heroes exemplified by the Trinity don Isaac Newton. This required a higher morality derived from within. We engage with these themes in the final three chapters.

CHAPTER ONE

TRINITY IN THE EARLY NINETEENTH CENTURY

> We have asserted that *we are light*, and that *we were darkness*, with equal eagerness and vehemence. In so doing we have been emboldened by the exclusiveness dominion which, during the last half century, Reason has usurpt over all our other faculties. The absolute supremacy of Reason, that is, of certain logical processes, in the concerns of mankind having once been acknowledged Reason too, according to the usual fate of despots, became a tyrant: nor was there any act of tyranny, however irrational, that she shrank from committing.
>
> Julius Hare, 1828

> "God," as said by the ancients, "works by geometry."
>
> William Whewell, 1833

The quest by those predominantly outside the Anglican Church for "rational" social, theological, and intellectual change was, despite the government's best efforts, gaining momentum throughout the 1810s. Tucked in the Cambridgeshire countryside, Trinity was certainly not insulated from such reform; the trick was how to accommodate transformation and yet retain the traditions of the past. It was not the role of Oxford and Cambridge to be at the vanguard of speculative knowledge but, rather, passing on the torch of known understanding. Oxford, for example, tried to ensure that only safe learning was prescribed by creating a New Examination System in 1801 to preserve orthodoxy rather than inquiry. It was hoped this would prevent the nurturing of a critical disposition among students that appeared to have caused so many of the problems in Continental Europe.[1]

It was within this context that the TC was tied, and they worked restlessly to try to harness history and truth with religion, custom, knowledge, and morality. This would involve a dialogue between an ever-confident emerging rational belief that the whole of the material world could be mastered and understood without God and the traditional view that man would always be subordinate to God's governance. What place was there for spirit, inspiration, morality, and divine knowledge in a fast-colonizing, human-centric, competitive, selfish, material, and "tyrannical" democratic world? What was the future of traditional social authority and the centrality of the Church of England within it? Faith was under scrutiny and attack like never before. Human improvement, for the TC, came about through a long period of history in which religion and morality, rather than political reform and material gain, were the true path to social rule.[2] Fear of the French and that country's revolution pervaded everything. Sedgwick wrote to his sister in 1804 that "England looked like a great camp," characterized by arming and drilling, even within the university.[3]

Other contemporary Trinity students were also engaged with such questions, including the pensioner Kenelm Henry Digby, who later toured Gothic churches with Whewell; William Sidney Walker, a sizar from Eton, who became a well-regarded Greek scholar and philologist; a pensioner from Paddington in West London and later prominent premillennial theologian, Edward Bishop Elliott; Hare and Thirlwall's two friends from their old school Charterhouse, George Waddington and Horace Waddington, who both subsequently became fellows of Trinity; and last, Thomas Thorp, a pensioner from Yorkshire who became president of the Cambridge Union in 1818 (Whewell had been president in 1817 and Thirlwall secretary). By this date student numbers at Cambridge had significantly increased, leading to the appointment of two proproctors to help the two proctors maintain university discipline.[4]

The following year Thorp won the chancellor's classical medal and became a fellow of Trinity and, in 1822, was appointed to an office in the college. He confessed to Whewell, "I should have hardly considered myself a fit person to lecture in Trinity College (tho' good enough perhaps for St. John's & I trust not bad enough for Sidney or Claire), had not the offer been actually made plump to me."[5] In 1833 he gained a Trinity tutorship and three years later was appointed archdeacon and chancellor of Bristol. He would go on to become Whewell's right-hand man as vice-master of the college in 1843. Intellectually, however, it seems Thorp "was widely regarded as a bumbler of little intellectual distinction." He was described by a former Trinity fellow, George Long, "as "the greatest noddy he ever knew."[6] Long was professor of Greek (1828–1831) and subsequently Latin (1842–1846) at University College, London. He spoke very highly of the talents of Sedgwick to James Gooden, a new student at Trinity, but "not so favourably of those of Whewell, who with vast arguments, it seems, is defective in arrangement and logical conclusions."[7]

Other contemporary members of Trinity were the Sedbergh-educated George Peacock; King Edward VI–schooled in Totnes and home-educated Charles Babbage; Richard Sheepshanks, who traveled to the European Continent with Whewell; and the Colchester Royal–schooled George Airy, who made pendulum readings with Whewell down in a coal shaft in Cornwall.[8] One should also mention the briefly Eton-educated John Herschel, who, although a student of St. John's College, Cambridge, had a great deal of contact with these Trinity undergraduates. However, these luminaries were in varying degrees peripheral to the TC.

WHEWELL, THE TC, AND THE PLACE OF MATHEMATICS

The Mathematical Tripos dominated education at Cambridge. But what type of mathematics should be taught? As we shall see, the introduction of French algebraic analysis was disturbing to several Cambridge students and dons. For outright hostility no one seemed so alien as the wealthy banker's son Babbage, who migrated from Trinity to Peterhouse the year "Billy Whistle" (pronouncing Whewell sounded like a whistle) arrived. Babbage came to represent the antithesis of the TC outlook.[9]

The TC's tolerance of this fallen Trinity student would greatly erode over time. Already in 1810 Thirlwall had proclaimed, "Never until genius may be purchased by industry will poetry become mechanical." There was no room for feelings or a soul in a mechanical (Babbagian) brain.[10] On this Wordsworthian and Coleridgean reading, deep emotions and intense feelings were best displayed through poetry, with divine sensibility the foundation to genius. Here imagination reflected God's creation, with language, especially poetry, the definitive act of this process. The better the poetry the closer you were to the original language and therefore a clearer notion of ideas. In other words, nearer to God.[11]

Related sentiments had been conveyed to Whewell by Rose in 1822. He despised proponents of Scottish philosophy when it came to religious issues, which along with followers of Locke's system led to "scepticism as to the existence of a first Cause." Moreover, he fumed,

> I hate both alike as leading to the cultivation of the sensible part of man instead of his spiritual & teaching people to occupy themselves with the theory of differences & the knowledge of formulae, which if a man knew all about them that was or can be known, leave him just where he was not wiser but fuller—not with a more occupied understanding: *leaving much instead the*

> *same sort of thing as Babbage's calculating machine—Indeed I believe that the approximation of the operations of the mind & of machinery is a very fit occupation for the Mechanic school.* (my italics)[12]

The emerging mechanical world took us further and further from our natural, simple, rural, and hierarchical state; the laws of human nature could not be found via a nominal abstract or simply sensory science.[13] By contrast, the further one got from the human scarred landscape the closer one got to God; a growing knowledge of the outside world depended on a growing knowledge of the internal consciousness. Emotion was the link between the inside and outside world with the act of introspection achieved via the imagination.[14]

This belief had not always been the case. For example, a young and keen Whewell, freshly arrived at Trinity, was initially wooed into joining the Cambridge Analytical Society (1812–1814). This body was set up by two of Cambridge's leading student lights, Babbage and Herschel. It was principally created to challenge what some of its undergraduate members, at this moment in time, considered the dilapidated intellectual climate of the university.[15] The aim of the society was to import the most powerful techniques of mathematical analysis and replace the Newtonian fluxional notation with the d's of Leibnizian differentials. Its inspiration derived from predominantly French mathematicians such as Pierre-Simon Laplace, Sylvestre Lacroix, and especially the powerful analytical machine of Joseph-Louis Lagrange's algebraic calculus. Should Britain be importing and adopting these predominantly French developments?

Cambridge University was in a nervous flux over such matters and French mathematics got caught in vociferous debates over curriculum reform. Numerous Trinity students joined the newly established Cambridge Union Society in 1814 to debate, among other things, political questions. They met in the smoky gallery of the Red Lion Inn in Petty Cury. Hare raised eyebrows when he gave a paper entitled "On the Question of the Propriety of the War against France." In the fraught atmosphere of increasing sectarianism and political radicalism—such as the rise of "Orator" Henry Hunt's public platform and the radicalism of Thomas Spence—the elder dons at Cambridge continued to be very uncomfortable with such student political activism.[16]

In March 1817 the university's vice-chancellor and master of St. John's, Reverend James Wood, had had enough and instructed William Okes of Caius and his college colleague, William French, to go and shut the Union Society down.[17] Whewell described Wood as "a damned vice Chancellor," Okes as characterized by a "round idiot head & turkey cock breasts," and French as being "inflexibly meek" with a "proctorial smile."[18] Wood and his colleagues were probably inspired by the 1817 Seditious Meetings Act just passed by Parliament. This outlawed all meetings of more than fifty people organized to discuss any grievance concerning the Church and state.[19]

Despite his swaggering remarks Whewell would soon be as conservative and orthodox as those he here condemned. This can be seen intellectually in his reversal of views toward French algebraic analysis and his embracing of traditional geometry. It was only Euclidian geometry that objectively revealed the properties of space. An empirical understanding of knowledge, as Joan Richards has shown, had clear problems with geometrical truth. As she writes, "An empirical theory of knowledge would suggest that valid knowledge is bounded by experience, and yet we have no experience of lines produced indefinitely in the far reaches of space. Euclid's definition of parallel lines asserted knowledge of how these lines would behave, however. Universally acknowledged to be true, such geometrical statements posed huge problems for nineteenth-century empirical epistemologists." Whewell saw such geometrical explanations of space as proof that knowledge was via divine fundamental ideas innate in our minds.[20]

Trinity's hero, Isaac Newton, writing in the late 1710s, had preferred presenting his work in geometry rather than algebra, "I was writing [*The Principia*] for Philosophers steeped in the elements of geometry, and putting down geometrically demonstrated bases for physical science." Since then, particularly on the European Continent, analysis had long been seen as superior to geometry, since algebraic proofs revealed and made visible the logical route to which a conclusion was made. By contrast, geometrical proofs involved all sorts of indirect and supplementary formations to reveal a conclusion. However, this was the point for the TC: it taught the actual intellectual path.[21]

The brief period of the Analytical Society's existence was an important moment in the adoption of differential calculus within English mathematics.[22] It was an attempt to move from a Newtonian emphasis on geometry and embrace Continental European approaches. Even Rose, who would go on to be the most anti-analytical philosopher of the TC, at this stage took a strong interest in mathematical algebraic analysis. Indeed, although it never appeared, he worked on a translation of Lacroix's mathematical work with Whewell in 1817—a year after the latter became second wrangler. Whewell was pipped to the top by the Westminster School–educated Edward Jacob of Caius and the son of the political economist William Jacob.[23] Whewell described him as "a very pleasant as well as a clever man, and I had as soon be beaten by him as by anybody else."[24]

Beside the main instigators of the Analytical Society, Babbage and Herschel, other notable members included Edward Bromhead of Caius (formerly educated at Glasgow University) and "Gentleman" Peacock.[25] All these figures, in varying degrees except Bromhead, went on to play important roles in establishing the Cambridge Philosophical Society (1819), the Astronomical Society of London (1820), and the British Association for the Advancement of Science in 1831 (although not Herschel in this case). During their time at Cambridge—Babbage, Herschel, and Bromhead veered toward what would have been considered politically radical views, while Peacock was a moderate Whig and Whewell, by the end of the 1810s, an all-out Tory. It is quite wrong to imply, as some historians do, that Babbage, Herschel, Whewell, and Jones formed a coherent intellectual group unified in their views toward science and reform. As we shall see, the situation was very different.[26]

Babbage's and Herschel's vision of the world must have made Whewell and Jones feel quite queasy in their chats during breakfast—particularly in their final undergraduate days. The one thing on which they were all united was the importance of induction and the collection of trustworthy observations—be they historical, economic, social, astronomical, or philological. The difference arose from the way those data were interpreted. Babbage and Herschel were moving, via algebraic analysis, to a human-centric rational symbolic logic. The brain was a machine perfectly formed to process and interpret empirical data. Both thus emphasized experience and a version of Locke's theory of a tabula rasa. Operating on algebraic symbols, they believed, put the naked workings of the brain on show. The brain could be rendered visible and available to empirical scrutiny and therefore open to the possibility of being built into a machine; it was no longer a mystical and sacred sphere. This was a quest Babbage would devote the rest of his life to and was the antithesis of the TC's view.[27] Here Babbage and Herschel were also, partly, drawing on the semiotic philosophy of the Frenchman Étienne Bonnot de Condillac and his impact on Lagrange, Laplace, and Lacroix. Condillac wrote in 1780 that "the technique of reasoning is thus the same in all the sciences. As, in mathematics, we state the problem by translating it into algebra; so in the other sciences, we state it by translating it into the simplest expression possible."[28]

For Newton at least, like his Trinity inheritors, the ancient geometrical techniques ensured that demonstration always applied to something real and revealed the complicated intellectual inductive journey to a theorem. This was not via some mechanical logic but a hard intellectual slog that, not infrequently, moved in unpredictable ways. By contrast, algebraic analysis would move into the abstract and only return to the original topic once a conclusion was met. The mathematics taught at Cambridge had developed within an environment designed to produce clergymen for the Church of England; in that sense geometry was as much a national institution as the Church. Coleridge understood and emphasized this.[29] Mathematical study was promoted as a means of reinforcing Christian faith rather than of discovering new facts via pure human reason; it was to train the mind to understand established knowledge (such as Newtonian gravitational astronomy) and to appreciate the evidence of the natural record. More important, however, were the teachings of Scripture and unfolding *all* our faculties to find one's soul, to gain a glimpse of the "Supreme Mind."

To adopt French abstract analytical mathematics would lead the mind into fantasy and, eventually, possibly atheism. The emphasis on algebraic mathematical rigor was bringing the discipline closer and closer to an insular human logic devoid of an active God. Abstract analysis is a system that attempts to ground anything complex into a set of simpler elements. By the mid-eighteenth century "simple analysis" was being employed to solve problems by reducing them to equations. Most analysts on the Continent were now linking calculus to the formal concept of a function instead of to conceptions of geometry. Many came to view analysis as the application of algebra to geometry: proof was demonstrated through algebraic techniques rather than geometrically. The Newtonian fluxional calculus held motion to be essential to the method of fluxions and was rejected as non-algebraic. In Cambridge, as Harvey Becher has shown, all "of the textbooks relied upon geometric figures rather than algebraic abstractions, ratios and proportions rather than equations, and specific physical foundations rather than mathematical generalisation."[30] Lines were real and mathematical symbols human invention.

Pure mathematics was not part of a Cambridge education; rather, the end of the Mathematical Tripos was Newton's *Principia*. Those advocating the algebraic analytical approach were concerned with grounding a perceived rational human logic and not with its application to natural philosophy. In other words, making the outside world, and indeed human morality, fit this logic. As Babbage and Herschel declared in their "Plan of a New Society" (1813), their aim was to ensure "physical science keep pace with the progress of analysis," and "will endeavour to introduce more generally the foreign notation in Analytical science regarding geometry and geometrical demonstration, as contrary to its ultimate objects."[31] This algebraic analysis was propagated in the form of Lagrange's formulation of the differential calculus based on the Taylor series.

Babbage and Herschel placed a great deal of emphasis on notation as the grammar of a universal language and key to making visible both nature and the mind. They were on a quest to find a mental technology that was universal, efficient, visible, and consequently immune to ambiguity and corruption—a trait they believed characterized the old paternal world of interests propped up by the Established Church and faith in the veracity of Scripture. Ultimately the latter was derived, they believed, from murky human testimony rather than mechanistic human clarity. Following Condillac, Babbage, in a paper entitled "On the Influence of Signs in Mathematical Reasoning," wrote, "The multitude of significations which attach to many of the words that compose our ordinary language, is a disadvantage which is completely removed from analysis . . . nothing appears so likely to succeed as a revision of the language in which all the results of science are expressed."[32] The TC would have been appalled by such claims.

Babbage was consistent in his very different TC view—from his "Act" the year before sitting the Tripos, when he defended his provocative statement, "God is a material agent," to later attempting to build thinking into a machine. Babbage's moderator at his "Act" was Reverend Thomas Jephson of St. John's, who branded him blasphemous and sent him down.[33] As such he was no longer able to take his degree with honors. Herschel too was far more of this latter persuasion.[34] Babbage was overjoyed when Herschel returned to St. John's as a subtutor and examiner in mathematics in 1815. He wrote, "what a glorious opportunity you have of spreading the true faith. . . . I hope you purge away the diagrams [geometry] which like cobwebs have obstructed their progress in the paths of truth." The Cambridge establishment was furious when Peacock introduced "d's into the papers" in 1817: Wood, the Sidney Sussex Fellow and Plumian Professor of Astronomy and Experimental Philosophy; Samuel Vince, the Trinity fellow and Lowndean Professor of Astronomy and Geometry; William Lax; and the master of Queens and Lucasian Professor of Mathematics Isaac Milner "were very angry and threatened to protest against analytic and French mathematics." Whewell likewise was very uncomfortable; Peacock's exam questions, he told Herschel, had "stripped his analysis of its applications and turned it naked among them."[35]

It was not that algebraic analysis was redundant; it could be a useful tool, but making it the sole basis of human reason made it dangerous and devoid of God. It certainly was not an instrument for the discovery of new knowledge. True reason required and derived through true divine faith. During the Napoleonic Wars, one writer in the *Eclectic Review* (1808) raged, "The writings of some eminent French mathematicians abound in infidel principles. Our elder men of science, we hope, are for the most part of too sober a cast to be injured by these principles; but we tremble for the fate of the young, and should rejoice to think it probable that these remarks might prevent one such student from abandoning the aids of religion, and venturing himself on the stormy ocean of life, without rudder, without compass, and without hope."[36] It was precisely this Anglican religious navigation that by the late 1810s the TC sought to preserve.

These were anxious times. It was not just the curriculum that was disturbing the dons but the moral behavior of the students—especially the arrogant and elite fellow-commoners running riot throughout the town and surrounding area. Milner was greatly upset by the "breaking of lamps and windows, shouting and roaring, blowing of horns, galloping up and down the streets on horseback or in carriages, fighting and mobbing in the town and neighbouring villages; in the daytime breaking down fences and riding over cornfields, then lastly, in the night frequenting houses of ill-fame, resisting the lawful authorities and often putting the peaceable inhabitants of the town into great alarm."[37] This helped prevent the introduction of gas lighting in Cambridge, for fear the students would simply go around smashing the lamps. However, it also meant privileged students could meet desperate prostitutes safely in the dark, before the latter returned to their run-down homes in nearby Barnwell; the contrast was stark.[38] Cambridge was becoming a town of thuggery, anarchy, and a den of immorality. The situation for many was symptomatic of the nation at large and needed a solution. However, it led to an angry and passionate debate concerning the right answer: rational or spiritual reform?

Herschel and Babbage were seeking to reform Britain's prevailing social hierarchy and institutions in accordance with rational human principles, emancipated from tradition, the Church, and the continuity of history—not so different, however it was clothed, from the dictates of the French Revolution. For the TC and countless others, this separation from the concrete past made abstract studies unnatural and politically subversive. For men like Babbage and Herschel human culture progressed through the logic of human reason, while for the TC an organic/spiritual culture preceded logical reason; the social order antedated the human intellect and thus defined the conditions of human interaction. The TC still retained a belief that human understanding had become corrupted after the Fall, and the only way to obtain a glimmer of truth was to access the divine ideas that were still within.[39] Our knowledge may have increased, but our intelligence was no better. Our objective was preparing for the next, not this, world.[40]

By contrast, Herschel wrote in his undated notes on "Great Physical Theories" that "the history of science in all its branches . . . shews that every great accession to theoretical knowledge has uniformly been followed by a *new practice*, and by the abandonment of ancient methods as comparatively *inefficient* and uneconomical"; both intelligence and knowledge progressed. Herschel, like Babbage, embraced the new urban-industrial world and saw the need to parallel this in human intellectual development, that is, the material process of reasoning.[41] However, with no spiritual safeguard this human-centric faith simply left a utilitarian quest for material gain.

As in industry, both Babbage and Herschel were obsessed with the conservation of mental labor and the increase in cerebral speed and believed this was possible via an algebra-based symbolic language. Babbage wrote, "The power which we possess by the aid of symbols of compressing into a small compass the several steps of a chain of reasoning, whilst it contributes greatly to abridge the time which our enquiries would otherwise occupy, in difficult cases influences the accuracy of our conclusions: for from the distance which is sometimes interposed between the beginning and the end of a chain of reasoning, although the separate parts are sufficiently clear, the whole is often obscure. *This observation furnishes another ground for the preference of algebraical over geometrical reasoning*" (my italics). By contrast for men like Whewell it simply effaced the true path to the conclusion and ultimately to the divine base to discovery. However, Babbage was evangelical in his faith in human reasoning. It was the rules of algebraic analysis, human logic, that divulged and discovered new knowledge. This was not necessarily anti-theological; rather, God had gone but had left the logical reason in humans to interpret the unfolding of divine design. In the beginning was not the word but the human faculty of abstraction. As Herschel reflected, "It is not truth that we have in *the abstract*—that we must find—but we have the *faculty of abstraction* to enable us to find it." Symbols did not depend on an external referent for their meaning, but simply on their initial relationship within the abstract system. This was faith in pure human logic. Whereas a mechanistic rationality made the observations fit algebraic analysis, geometry waited until the observations formed or recognized a clear idea. In other words, for Herschel and Babbage, the clunking of the human brain alone—without the need for divine innate ideas—was the way knowledge was discovered. Herschel made it clear in his essay "Mathematics" for the *Encyclopaedia Metropolitana* (1830) that, as such, a mathematical approach required "no assistance from inductive observation, and very little from the evidence of our senses." It was through the brain's correct algebraic logic that observation could be guided and confirmed.[42]

Perhaps the person who came to embody this perspective the most in nineteenth-century Britain was the shoemaker's son George Boole, who was brought up as a member of the Church of England in Lincoln but became a Unitarian and opponent of the 39 Articles. He was also a Chartist sympathizer.[43] He went on to develop the kind of symbolic algebra that Babbage had been promoting during the 1820s and was a fan of Babbage's unofficial *Ninth Bridgewater Treatise* (1837). In addition, he visited Babbage's house much later to hear how the Difference Engine worked. It was during this decade, the 1830s, that Babbage developed his idea of an Analytical Machine, an engine capable of applying a universal algebra that could manipulate both numbers and symbols.[44]

Boole's method was a calculus of deductive reasoning; logical propositions could be deduced by symbolical algebra.[45] Although he was patronized by a former member of the Analytical Society, Bromhead, the TC would not have paid him much attention. He was, after all, simply a peripheral teacher working in the north, including the Lincoln Mechanics Institute. Nonetheless, he was aided by the Trinity mathematician D. F. Gregory, the Trinity assistant tutor Arthur Cayley, and the fallen Trinity student and then professor of mathematics at University College, London, Augustus De Morgan. Thanks especially to their patronage Boole gained the first chair in mathematics at Queens College, Cork, in 1849.[46]

It was at the Lincoln Mechanics Institute in 1835 that Boole gave his first public lecture, which was on Isaac Newton. Despite his general celebration of Newton's work Boole slammed his mathematical approach: "There was yet another disadvantage attaching to the whole of Newton's physical inquiries which, though it gave rise to the most sublime applications of Geometry, must yet be considered as having presented an insurmountable barrier to his progress—the want of an appropriate notation for expressing the conditions of a dynamical problem, and the general principles by which its solution must be obtained. By the labours of Lagrange, the motions of a disturbed planet are reduced with all their complication and variety to a pure mathematical question."[47] By contrast Newton's contemporary, Leibniz, was far less dependent on diagrams and geometric models. For men like the TC these would have been offensive and critical remarks.

Had they considered him important, the TC would have abhorred Boole's attempt to reduce the workings of the mind to a mathematically grounded logic. In 1847 Boole wrote, "The mathematics we have to construct are the mathematics of human intellect." He politely dismissed the TC's concern that the application of symbolic analysis weakened the reasoning faculty. On the contrary, like Babbage and Herschel, he argued its application was vital to scientific discovery.[48] He greatly expanded this argument in 1854 in his *Laws of Thought*. In Painite fashion he reminded his readers "that it is the business of science not to create laws, but to discover them." This was done via human applied symbolic analysis. "Again, whatever other object this treatise may accomplish, it is presumed that it will leave no doubt as to the existence of a system of abstract principles and of methods founded upon those principles, by which any collective body of social data may be made to yield, whatever information they implicitly involve." The "ultimate laws of thought are mathematical in their form."[49] It was such a conclusion that the TC had long feared.

Already for the TC by the close of the 1810s, French algebraic analysis was particularly associated with the unnatural events in France; it was the horrifying result of human intellect set free from all Christian theology and social authority. It was not that some Cambridge dons and students did not understand it, but they disliked its implications (total faith in human abstract reasoning). In many ways, it represented the personification of man's corrupted reason since the Fall. The result, they claimed, was a product of human fantasy and folly, cut loose from established religion, tradition, and reality. The abstract nature of pure algebraic analysis seemingly allowed humans to wander into false terrain through the meaningless manipulation of symbols; this was a world devoid of an active deity and need for God. It related to a human-centric, utilitarian, and industrial view of the mind; the world, both physically and socially, was being made to fit algebraic analysis rather than the opposite—a world based on false human mechanistic reason alone and not God's living organic template. Such algebraic analysis was simply concerned with the logic of combining things, rather than defining the things to be combined. This was an artificial intelligence making an artificial world. Of course, there were many who tried to refute this, but the differences tended to be unclear and unconvincing.[50]

There is no doubt the social disintegration of the late 1810s was turning Whewell against reform and into an orthodox Tory. This is the key to understanding his vehement switch from his pro–Analytical Society days to one of anti-algebraic analysis. Indeed, he held ancient Greek Euclidian geometry as the ideal subject for a university education. Bishop George Berkeley had likewise, much earlier in his *Analyst* (1734), argued strongly for the preeminence of geometry as divine rational thought. It taught one to think exactly, he argued, and could be transferred to other areas to find truth, a point Whewell made when he later worked on forging a system of morality. An analogous debate also centering on the place of human logic was happening elsewhere. The leading Oxford

Tractarian, John Henry Newman, saw human logic as artificial and, as Henry Chadwick puts it, "sharing with pure mathematics the danger of being exposed to corruption as soon as they are applied inappropriately to actual experience of life."[51]

As knowledge was breaking down so was social authority and morality. Parts of the country, not quite yet Cambridge but certainly parts of Cambridgeshire, were electric with reform; mass political meetings spearheaded by radicals, such as Hunt, were in full swing. The popular mass platform movement wanted the reduction of indirect taxes on everyday items, parliamentary reform, and the end of the Corn Laws.[52] Already in 1816 Whewell was beginning to get concerned about the breakdown of traditional social rule. For example, he became anxious about riots going on in nearby Littleport and Ely. The commotions were, like elsewhere in the country following the Napoleonic Wars, triggered by high unemployment. An uprising broke out in Littleport after some folk drinking in the Globe Inn set out to intimidate the local wealthy inhabitants. The situation escalated and property was destroyed, coupled with aggressive demands for money. The result was the Ely militia, supported by a regiment of the Royal Dragoons, crushing the rebellion. During the main clash at the George and Dragon in Littleport, one protester was killed via a shot to the head and one soldier was injured. Eighty-two demonstrators were captured, resulting in six being acquitted, ten discharged, thirty-six bailed, twenty-four condemned with nineteen of them having their sentences commuted. Of these nineteen protestors (eighteen men and one woman), nine were transported to Australia for fourteen years, and a further ten for seven years. The final five men were executed in Ely on June 16, 1816, with a plaque subsequently erected at St. Mary's Church in the city as a memorial to their deaths.[53] As a precaution Cambridge swore in many new special constables to patrol the town.[54]

Sheepshanks wrote to Whewell in March 1819 describing a rally he experienced in York. He estimated there were "from five hundred to a thousand decent people there & from fifteen to twenty thousand radicals nearly all from our good town & neighbourhood." He feared for the future of the country and wished his "family & property were quietly out of the way."[55] All the TC would be further dismayed in August that year in the wake of the Peterloo Massacre at St. Peter's Field, Manchester, when the cavalry charged an estimated crowd of sixty to eighty thousand people. No doubt the icing on the cake for Whewell and his now complete suspicion of political, institutional, and intellectual reform came with the Cato Street Conspiracy in 1820 and the attempt to murder the prime minister, Lord Liverpool, and his cabinet ministers. The French Revolution and everything associated with it—especially its human-based rational philosophy—were dangerous, unnatural, and destructive; everything had to be done to prevent this disease penetrating Britain's institutions, education, and establishments. The latter—the monarchy, Parliament, and the Church—tended to evoke a sense of Britishness, certainly Englishness, more than the ethnicity that characterized the rise in Continental European nationalism.[56]

Whewell later found unequivocal support for his arguments on education, particularly the role of mathematics, from the Eton- and Christ Church–educated constitutional historian Henry Hallam, who congratulated him on his support of geometry and propagation of established divine fundamental ideas. However, Hallam had to confess being pleasantly surprised that someone in Whewell's position should still prefer geometry to algebra as the foundation of permanent studies in mathematics. He totally agreed with Whewell's conclusions regarding geometry—"that it is more [divinely] logical & deductive, & that it makes the properties of the external world more distinctly before the eye."[57] This emphasis on geometry kept reasoning within the realm of a traditional Anglican education, in which the mind remained spiritual and sacrosanct rather than rule-bound and disciplined by a false human logic.

Whewell's views were now not only equivalent to older British critiques of algebraic analysis but written in the same way. Consider the late eighteenth-century Scottish philosopher of natural philosophy John Robison. During this period he was at the vanguard of Edinburgh University's assault on any philosophers employed by the institution who smacked of Jacobinism.[58] In 1797 he published his paranoid *Proofs of a Conspiracy against All the Religions and Governments of Europe*. Seven years later his *Elements of Mechanical Philosophy* appeared. The Scottish philosopher of logic and metaphysics William Hamilton later wrote that Robison "preferred the ancient method of studying pure geometry, and even felt a dislike for the Cartesian method of substituting symbols for the operations of the mind, and still more was he disgusted with the substituting of symbols for the very objects of discussion, for lines, surfaces, solids, and their affection. . . . And *he came at last to consider algebraic analysis little better than a kind of mechanical knack, in which we proceed without ideas of any kind, and obtain a result without meaning and without being conscious of any process of reasoning*" (my italics).[59] Whewell now clearly shared Robison's view, and such a stance was embedded among much of the philosophical establishment at the time. The leading English savant, Thomas Young, also condemned the French, industrial, and mechanical nature of modern analysis:

> The strong inclination which has been shown, especially on the continent, to prefer the algebraical to the geometrical form of representation, is a suf-

ficient proof, that instead of endeavouring to strengthen and enlighten the reasoning faculties, by accustoming them to such a consecutive train of argument as can be fully conceived by the mind, and represented with all its links by the recollection, they have been desirous of sparing themselves as much as possible the pains of thought and labour, by a kind of mechanical abridgement, which at least only serves the office of a book of tables in facilitating computations.[60]

And later in 1810 he wrote: "French mathematicians too, Lagrange not excepted, have fallen into the illogical habit of generalizing too soon, of suspending conviction upon some remote analogy, instead of founding it upon irrefragable evidence . . . as incontrovertible, and as obvious, as those which served for the foundation of the first elements of geometry."[61] This could have been straight out of Whewell's later writings on the subject.[62]

Jones later rejoiced at Whewell's "effort to purify & amend the mathematical training of the place. I have long been concerned that as a matter of training exclusive habits of symbolical reasoning are not merely useless but deleterious & I see very often instances of their bad efforts on men of very acute minds." Whewell also received glowing praise from his soon-to-be Oxford collaborator in the fight against the reforming university commissioners, Edward Hawkins, educated at Merchant Taylors School and now the provost of Oriel: "What I have read . . . has uniformly appeared to my mind correct. And I hope your opinion may become known here [Oxford] as well as Cambridge; for there has been some tendency here also to go on too rapidly into 'Progressive studies' and to give up Geometry for Analytics."[63] By the end of the 1840s Hawkins was pleased to tell Whewell that the revised Oxford curriculum was "much more like your plan" and carried by a large majority: "It is your school of Moral Science, or nearly so, but without Political Economy."[64] Whewell could not have been happier.

This was in direct contrast to reformers such as Babbage, who had a notion of intelligence, which one recent commentator has directly referred to as a factory.[65] God had legislated the world and left it to operate by itself—there was no miraculous divine intervention or fundamental ideas, only an occasional gear change. "God at Creation had appointed new sets of animals and plants to appear like clockwork throughout history—he had created the laws which produced them, rather than creating them directly."[66] This was not a massive leap, in fact, from Paley's evidence for divine creation.[67]

Babbage, like David Hume earlier, dismissed miracles.[68] In his unsolicited *Ninth Bridgewater Treatise*, he gave his alternative explanation:

First he asked, when do we become satisfied, through induction, that a series of numbers follow a law—after 500, 1000, or 50,000? He then gave a series of numbers that suddenly changed—at 100,010,002—from an arithmetical change of 1 at every operation to a series of triangular numbers, each multiplied by 10,000, which again changed at another number to a new law. Babbage claimed that a machine that has "embodied in its mechanical structure, that more general law of which all the observed laws were but isolated portions . . . displays far greater generality than that which demands, at every change in its law, the direct intervention of its contriver."[69] God had gone and the world was deterministic via the divine blueprint. This was a brilliant and powerful argument. Although not a new idea, it was given extra depth under Babbage in his description of the process and how it could be demonstrated via a machine. The Analytical Engine could be made to produce a series of natural numbers in regular order for a certain number of places and then switch to another law—and so on. Much the same "switch" happened, argued Babbage, when a caterpillar changed into a butterfly or a geological change occurred on the earth's surface. Both alterations, he claimed, were analogous to mechanical changes of law.[70] This was not Darwin's subsequent evolution via natural selection but prescribed mechanical determinism (although there was room for the two to be conflated—perhaps God had already decided selection). Despite Whewell's admiration for Herschel, the footprint of this celebrated St. John's astronomer, along with Babbage, was only too evident within the prevailing mathematical curriculum at Cambridge. For the TC reformers, this perspective radically and dangerously undermined their view of a liberal education and the sacred relationship between the Church and state.

The idea that the mind was a machine or factory was blasphemous and driven by a false, superficial interpretation of the intellectual rise of an industrial society. Whereas Babbage, and to a certain extent Herschel, craved mechanistic utilitarian institutional reform, the TC sought careful, slow, organic change that retained its past. Within this objective, language was intrinsic to permanent knowledge, and history was the tool to finding it. Crucially too, it separated man from animals. In what Richard Yeo terms Whewell's "metascientific project," he did not seek to ground rules of science, but rather to generalize through the history of science the pattern in which subjects were translated into scientific disciplines.[71] This was a divine, not human-centered, process.

The task of the Anglican universities was to inculcate students with those ideas that would teach them to take their places as leaders and not as critics of society and its established institutions. As a former senior wrangler from 1791 and fellow of Trinity, Daniel Mitford Peacock, warned in 1819, "There is in truth no one point on which the University should be more upon its guard, than against the introduction of mere algebraic or analytical speculations into its public examinations." Rather, "academical education should be strictly confined to subjects of real utility" (demonstrated things), unlike "the lucubrations of French analysts" (false things), which "have no immediate bearing upon philosophy" (real things).[72] Mathematics was becoming an independent discipline severed from the physical sciences and associated with a mechanistic logic emphasizing total faith in human reason.

Whewell's quest was now somehow, almost reluctantly, to retain mathematics as central to a Cambridge education but without creating a calculating clergy and reformist utilitarian students devoid of religiosity; they needed to be installed with an appreciation of the country's organic past and the nation's one Church and associated morality. His solution would be a traditional form of mathematics (most notably Euclidian geometry) and Newtonian mechanics—along with, virtually, an accompanying system of morality that was irretrievably tied to such a true divine tool and subsequent derived knowledge; both were predetermined by God in man's soul. At the heart of Whewell's pedagogical philosophy were divine "fundamental ideas" found within us but contingent on induction and imaginative guessing. This was the antithesis to a mechanical logical association of sense data—an approach most associated with Locke and later David Hartley and especially deductive French mathematicians. Whewell concluded in his *Thoughts on the Study of Mathematics at Cambridge* in 1835, "It is highly for the interest of the University that the over disposition to analytical generalizations should not be fostered," and the emphasis on algebra be replaced by the teaching of geometry, mechanics, and hydrostatics.[73] Euclid of Alexandria (mid-fourth to mid-third centuries BC) was far closer to the mind of God than modern mathematical analysts.

CAMBRIDGE AND THE ANGLICAN RESPONSE

For the TC both France and, increasingly, Britain had and were being corrupted by an artificial—human-centric—corrupted intelligence. Their solution was the need for the word of God to be spread more than ever to purify this false environment. However, it had to be the correct divine word. This was a widely shared aim. The evangelical chancellor of the Exchequer and Christ College, Oxford–educated Nicholas Vansittart passed an act granting £1,000,000 for the building of Anglican churches in perceived godless industrial areas in 1818; a further £500,000 was passed in 1824.[74] A key factor in the design of these new churches was for the congregation to clearly hear the service and, by the Victorian period, for the building to be preferably of Gothic design. One reason the poor frequently chose to worship at dissenting chapels was the more comfortable seating and less hierarchical social arrangement.[75] The Anglican pew system was widely disliked and more concerned with spectacle and smell than with the word. It was the Dissenters who had traditionally built preaching boxes, always in a classical idiom—although Dissenters did turn to Gothic in the 1840s as a sign, perhaps, that they were becoming respectable and hierarchical in their turn.[76]

The impetus behind building these new Anglican churches was the elite clerical and High Churchmen. Men such as the Etonian and King's graduate Charles Simeon, nicknamed "Chin Simeon" due to his elongated jaw, who dominated evangelical Anglican life in Cambridge as the vicar of Holy Trinity Church.[77] This Gothic Perpendicular church is on the intersection of Sidney Street with Market Street in the center of Cambridge. The TC would have been very aware of Simeon's presence and teachings.

Simeon also went on to be one of the founders of the Church Missionary Society in 1799, the London Society for Promoting Christianity amongst the Jews in 1809, and an adviser to the East India Company on their appointments of chaplains to India. Other important figures promoting church building were the members of the Hackney Phalanx Society (1805–1830). By 1824 some 218 new churches had been built. Between 1818 and 1843 the subdivision of parishes was increasingly replaced by the creation of much larger ones. In addition, between 1840 and 1876 a further 1,727 churches were built. Income for curates within these much larger parishes was predominantly via pew rents and fees from their parishioners—this was to ensure they did all they could to encourage attendance. The Church of England was desperately working throughout the century to retain its monopoly status.[78]

Building Anglican churches was one thing, but something also had to be done for a Church of England education. If mathematics was going to remain at the center of the Cambridge curriculum, all such textbooks were and should be dependent on geometric representation and, where applicable, on specific physical foundations rather than algebraic abstractions and mathematical generalization. A strict Anglican education held traditional geometry to be the means to train the mind to conform, and not go beyond what was known from prescription, experience, and applied to real things (God's world). It was, as we have seen, considered the most important tool for understanding the great Trinity natural philosopher, Newton, and his *Principia* (1687)—which was viewed as the exemplar of human understanding of God's natural world.[79]

It was within this anxious environment that the TC had formed a reading club. While Babbage sat in his college rooms and established a group concerned with importing French algebraic analysis, Hare sat in his digs and created a party around introducing German literature and discussing the beauty of the "Romantic" English Lake verse. Hare, Thirlwall, Rose, and Whewell spent many evenings studying German and learning an appreciation for the work of Romantic poets.[80] Rose and Hare particularly worked on Whewell to value Wordsworth.[81] Whewell, much to their delight, won the Chancellor's Gold Medal for an English poem with a Wordsworthian poem.[82] He waxed lyrical over the work of Schlegel (probably Friedrich who, along with his brother, August Schlegel, led the Jena Romantics and founded *The Athenaeum* in 1798).[83] Whewell was now confident about his mastery of German philosophy. He told Rose, "If you have not read the book I think you will find it will repay you for the perusal. Hare considers it as the ideal of criticism."[84] Clearly at this point Rose had not yet taken his anti-German turn. However, both French mathematical and German literary intellectual imports were, as we shall see, as controversial as each other. It should be remembered how many of the German philosophical Romantics had initially defended the French Revolution.

Whewell was increasingly worried about the state of religion. He told his sister that it was becoming too extremist, with the evangelicals having far too "violent notions about absolute predestination to salvation." Whewell was here falling into the common trap of identifying nineteenth-century evangelicalism with Calvinism. By contrast, at the other end of the spectrum Unitarians rejected everything and reduced all "to mere human reason."[85] Salvation could only come from a robust Church of England.

TRINITY AND THE HACKNEY PHALANX SOCIETY (1800–1830)

Whewell, Hare, and Thirlwall gained fellowships at Trinity. Rose obtained the curacy of Buxted, Surrey, and became increasingly connected to the Hackney Church Phalanx Society. Other members of this Anglican arch–High Church group included the Winchester and Trinity don Christopher Wordsworth, whom Rose went on to appoint as a rector at Buxted; while two Trinity fellows, namely the Charterhouse-educated Regius Professor of Greek James Henry Monk, and later the Bury-based King Edward VI school (St. Edmunds)–educated bishop of London Charles James Blomfield—were also associated with the society. Monk and Blomfield were preferred into the episcopate by Prime Minister Liverpool.[86] They were further joined by two other Trinity graduates, the successful lawyer and the Cheltenham-based, King Edward VI school–educated Nicholas Conyngham Tindal and the Etonian Charles Manners-Sutton.

Manners-Sutton was the son of the archbishop of Canterbury and, while at Trinity, had Christopher Wordsworth as his private tutor. Manners-Sutton subsequently awarded his teaching mentor by appointing him his chaplain when he became speaker of the House of Commons in 1817. He was also instrumental in ensuring Lord Liverpool appointed Wordsworth as the new master of Trinity in 1820 after the death of Bishop William Lort Mansel.[87] Other Cambridge members of the Phalanx included the Peterhouse student Henry Handley Norris and the Etonian, Corpus Christi graduate, and onetime domestic chaplain to Manners-Sutton's father at Canterbury George D'Oyly, who was also later appointed a rector at Buxted by Rose after D'Oyly's father had died in the position. Finally, two further Cambridge acolytes were associated with the society and both came from King's. First, the Etonian John Lonsdale, who was also a onetime chaplain to Manners-Sutton's archbishop father and who subsequently succeeded Rose as principal of King's College London in 1838; and second, the Etonian, lawyer, and later judge John Patteson.[88]

The Phalanx was prominent in the control of the High Church review *British Critic* and created the periodical *Christian Remembrancer* in 1819. Among many things, they strongly opposed the Lancastrian system of education for the poor set up by the pioneering Quaker Joseph Lancaster. In the educational fight, they mobilized their enormous wealth and power to form the National Society for Promoting the Education of the Poor in the Principles of the Established Church as an alternative in 1811. President of the society was Archbishop Manners-Sutton, with the prince regent as a prominent patron. They raised £16,000 in the first year, and by 1815 every diocese was in contact with the society; by 1824 there were some three thousand schools teaching half a million children the doctrine of the Church of England.[89]

THE NEW CLASSICAL TRIPOS AND THE CAMBRIDGE APOSTLE SOCIETY

On November 13, 1820, some of Cambridge, but not any of the colleges, was illuminated by flames. Riots broke out because of the abandonment of the Pains and Penalties Bill that was introduced by Parliament, at the request of King George IV, to dissolve his marriage to Caroline of Brunswick. She was a very popular figure and this act deprived her of the title of queen of Great Britain and Ireland. There was fierce town-and-gown fighting around the area of Market Hill, with some 250 special constables sworn in to supress the disturbances.[90] Tensions within Cambridge were volatile and high.

At this point the appointment of the enthusiastic classicist Christopher Wordsworth as master of Trinity and shortly later vice-chancellor was great news for the TC. Student numbers at the university expanded by about a third during the 1820s. As well as being important in helping to introduce a new Classical Tripos to the Cambridge curriculum in 1822, Christopher Wordsworth played a leading role in expanding Trinity and the building of New Court helped by a donation from the king of £1,000. At its opening the speaker of the House of Commons, Sutton, laid the first stone.[91]

Thanks to Whewell and his close relationship with Wordsworth, Hare was appointed as classical lecturer at Trinity in 1822 and Thirlwall four years later. The key architects of the new Classical Tripos were the Regius Professor of Greek, Monk, and the master of Christ's, Charles Kaye. Wordsworth took a back seat after suspicion about his intentions were expressed by Trinity's main rival college, St. John's. The result was a watered-down voluntary examination that excluded Latin and Greek composition and was available only to those who obtained honors in the Mathematical Tripos. At Oxford the opposite was the case. Nonetheless, the earlier TC reading club of Hare, Thirlwall, Whewell, and Rose had achieved their objective of spreading their passion for classical antiquity among Cambridge students.[92] Christopher Stray, interestingly, speculates that the introduction of the Classical Tripos could have been a way of countering the growing emphasis on algebraic analysis in the Mathematical Tripos. This would certainly explain the suspicion toward Wordsworth from the leading mathematical college at St. John's.[93]

Whewell was thrilled by Thirlwall's appointment in 1826: "This is good in every way."[94] Throughout this time, they were all regular guests of Wordsworth at Trinity Lodge.[95] Now the TC had a real chance of making an impact on the college's instruction. As Hare reflected to Whewell on their friend's selection, "I hope now a little more confidently, that we shall be able to do something in Cambridge for classical literature."[96] Drawing on German philosophy and hermeneutic techniques, the study of the Bible and ancient texts would become much more prominent. This onetime undergraduate reading group, minus Rose, now formed a formidable teaching team led by Whewell at Trinity, although the latter would remain primarily committed to teaching mathematics, with the classical side left mainly to Hare and Thirlwall.

Before taking his fellowship, Hare took a trip to Italy where he met Thirlwall's philological guru, Karl Josias von Bunsen, who, along with the writings of two former Cambridge University students, William

Wordsworth and Coleridge, greatly inspired him. Thirlwall and Hare's best-known students were the theologian Frederick Denison Maurice and the Scottish writer John Sterling, who came down from the University of Glasgow; both, however, migrated from Trinity to Trinity Hall to further their study of law.[97] The former was also an early member of the Cambridge Conversazione Society, the so-called Cambridge Apostles (est. 1820), an elite group of twelve who met weekly, and still do, to discuss moral and social topics along with developing their literary interests, in particular poetry.[98] The apostles took over the editorship of a magazine entitled the *Athenaeum* in 1828, which became a Trinity organ in the fight against utilitarian materialism.[99] A specific Anglo-hybrid version of German idealism, which underlined ideas connecting humankind with the material world, was emphasized; matter was impregnated with mind. This focus by Whewell, Hare, and Thirlwall on divine intuition, it was hoped, would undermine the erosive materialistic philosophy at large.

Bunsen, a student of the great German scholar Barthold Georg Niebuhr, in some sense had defined the TC project in 1817. For example, he wrote to his sister that year, "The consciousness of God in the mind of man, and that which in and through that consciousness, He has accomplished, especially in language and religion, this was from the earliest point before my mind." The way to revive religion was to reveal God's ongoing revelation throughout history via the study of philology, the imagination, and intense inner introspection.[100] This also characterized Coleridge's later work—the history of language was a history of consciousness. Thus to understand the meaning of a word from the past provided an insight into that historical context.[101] In 1828 Hare and Thirlwall translated the first volume of Niebuhr's *History of Rome*, followed three years later by the second volume. The translation became an important text at both Cambridge and Oxford and paved the way for critical studies of all ancient history.[102]

At the heart of the TC was the study of language. This was the real archaeology of humankind. It was not some sensory reflex but a creative act of the mind that, ultimately, stemmed from God; there was and could not therefore be any such thing as human evolution. No other species of God's creation spoke a language, which was surely proof of man's divine and special nature. In one of his last works translated into English posthumously, *God in History; or, The Progress of Man's Faith in the Moral Order of the World*, Bunsen wrote in quite a Wordsworthian fashion, "The power of the mind which enables us to see the genius in the individual, the whole in the many, and to form a word by connecting a subject with a predicate, is essentially the same which leads man to find God in the

Universe and the Universe in God. Language and religion are the two poles of our consciousness, mutually presupposing each other."[103] This was the vital quest of the TC and their underlining of imagination with poetry and the languages of antiquity. Language revealed divine providence. Following Coleridge especially, language organically developed like branches and leaves from an original trunk.[104] To discover a word's origin (etymology) brought you closer to its uncorrupted meaning.[105]

As well as becoming a leader of Protestantism in Europe, Bunsen had close social links to Cambridge, and especially Trinity, after marrying Frances Waddington, who was a cousin of Monk.[106] It was primarily the TC, especially Hare and Thirlwall, that initially set the pace in the study of ancient languages in Britain. On returning to England from Rome, where he had met Bunsen, Thirlwall continued his German studies with Hare and engaged with the quest to find the origin of the Bible. This led to his translation of Friedrich Schleiermacher's *Critical Essay on the Gospel of St. Luke* (1817) in 1825; this was a very methodological approach compared to traditional eighteenth-century ways of defending Christianity. Here the divinity of Christ was denied and instead presented as more of a symbol or representation of Christian faith. Religion was a product of feeling that came before cognition and volition; you did not have to abandon Christianity if you did not believe in miracles. Important too, he argued, was the need for an ethical approach to the natural sciences.[107] This, as we will see, came to bite Thirlwall hard in later life. It was through reading Schleiermacher's studies of Plato that John Stuart Mill started to question some of the principles of utilitarianism and find himself attracted to aspects of Thirlwall's and Hare's work.[108] The true human character, for Schleiermacher, came from the Church—a view that had earlier informed Friedrich von Hardenberg (Novalis). However, Schleiermacher's reconciliation of incompatibility between religion and science would, once again, soon dissolve into a clash between empiricism and evangelical theology.[109]

Hare and Thirlwall greatly informed the thinking of Maurice and Sterling, in particular their attraction to Bunsen, William Wordsworth, and Coleridge. Indeed, the latter two British poets were a primary resource for a great many during this period, including the Oxford Movement, and are both seen as important in the contemporary religious revival. They emphasized spontaneous natural feelings against rationalist standards.[110] Of the two, however, it was Coleridge's emphasis on the unseen rather than the aesthetic natural world that proved the most important. Coleridge was Paley's fiercest critic and drew particularly from the work of the "all-crushing" Kant along with other German idealists, such as Johann Gottlieb Fichte and Friedrich Schelling. The latter was important in both Germany and England in supplying a scaffolding for Romantic theory. Unlike these German philosophers, however, Coleridge's emphasis was fully on religion.[111] Maurice later reflected, "I, in a small society of which I was a member [the Cambridge Apostles], defended Coleridge's metaphysics and Wordsworth's poetry against the Utilitarian teaching."[112] Coleridge sought a single principle, via a universal Christian Church, that could fuse all his thoughts and, indeed, all minds. Such an ambition had the chance of bringing together all doctrines. This was the quest of the TC and disciples such as Maurice: universal truth came through universal theology.[113] An attempt in 1835 was made by Hare and Whewell to establish a Coleridge Essay prize at Cambridge; however, the vice-chancellor and master of Trinity objected. They claimed his early radical writings were still more well known than his later Christian philosophy.[114]

The shared sentiments of the TC can be found in Hare's sermon "The Children of Light" in 1828 and, along with his Oxford-educated brother, Augustus William Hare, their cowritten *Guesses at Truth* published the year before. Like Rose and, as we shall see, Whewell and Jones, human reason as the basis of epistemology was creating a climate of dangerous political reform and intellectual distortion, driving "man" further and further from God and traditional social authority. Human reason alone was imperfect and damaged faith—it produced a false understanding and morality. This was for the TC an age of anxiety. Hare traced this selfish materialist philosophy to Thomas Hobbes and its most pure manifestation in Benthamism. The latter movement, what Mill coined "Utilitarianism" and trust in the greatest number, led man away from truth and higher states of knowledge such as art, imagination and poetry.[115] The radical creed of progress challenged men like the TC and became associated with revolution and godlessness. As Duncan Forbes once put it, "'Civilisation' and the 'March of Mind' were assumptions which the Liberal Anglicans could not accept, because they dispensed with God's Providence."[116] Their lack of touch with everyday folk was breathless—but that was the world they frequented.

In "Children of the Light," Hare declared scientific education to be utterly "heartless" and that moral philosophy and religion should be at the center of an education. It was by nurturing reason in conjunction with imagination that the laws of God could be detected and one's moral duties correctly emerged. The faculty of human reasoning, to work, had to coexist with the faculty of the human imagination and human heart. Otherwise it imposed a false form on the world: "Indeed all intellectual wrong, when traced to its source, will be found to take its rise in moral wrong." By ignoring all the other faculties, it created an artificial world such as France. Hare concluded, "Only when they [all our faculties] are knit together, and working in unison and harmony, may we hope that the vision of Truth will descend upon them."[117]

In the *Guesses* the Hare brothers wrote, "The foundation of all happiness, temporal and eternal, is faith in the goodness, the righteousness, the mercy, and the love of God." Maurice felt the book would unfortunately not be popular, as the brothers "do not profess to be a system . . . but they think systematically." However, he was wrong, and the work went on to become a best seller. Along with Sterling, Maurice would gain a great deal of support and patronage throughout his career from his old teacher.[118] In the *Guesses*, the Hare brothers saved special mention for Bentham and Hobbes, the "chiefs of the selfish school" and false views: "Both have the same arrogant, overweening, contemptuous self-conceit. Both look with the same vulgar scorn on all the wisdom of former times, and of their own. Both deem they have a monopoly of all truth, and that whatever is not of their own manufacture is contraband."[119] There was a growing audience that delighted in these venomous remarks.

Trinity had now emerged at the forefront of promoting German thought and teaching ancient history, art, and institutions. The TC would have been thrilled when composition in Latin and Greek was added to the Classical Tripos in 1830.[120] Prominent students arising from the new Tripos included Whewell's favorite, the later poet laureate, Louth-based King Edward VI school– and Trinity-educated Alfred Tennyson; the Harrow and Trinity poet and later Anglican archbishop Richard Chenevix Trench; the Bury St. Edmunds (King Edward VI)– and Trinity-taught Apostle Society member, historian, and philologist John Mitchell Kemble and another apostle, the Harrow- and St. John's–taught Charles Merivale.[121] Merivale later reflected, "We began to think that we had a mission to enlighten the world upon things intellectual and spiritual. We held established principles, especially in poetry and metaphysics, and set up certain idols for our worship. Coleridge and Wordsworth were our principal divinities, and Hare and Thirlwall were regarded as their prophets; or rather in this celestial hierarchy I should have put Shakespeare at the top of all, and I should have found a lofty pedestal for Kant and Goethe [Johann Wolfgang von Goethe]."[122] Trench later reflected that his friends were "that gallant band of Platonico-Wordsworthian-Coleridgian-anti-Utilitarians."[123] Kemble almost got expelled from the university for describing Paley as a "miserable sophist" and condemning Locke for expressing "loathsome infidelity."[124]

Hare's good friend, Thomas Arnold, came close to equating poetry with religion—in contrast to a material world that had severed us from our feelings and morality. Or put another way, good theology is good poetry; it was the mirror to the soul. Indeed, it was Hare who introduced Arnold to Niebuhr's *History of Rome*.[125] However, this was going against a growing materialist view that poetry was illusionary and science, via human senses and logical reason, was real; the latter was progress and the former backward.

THE PROBLEM WITH METAPHYSICS

Not all the TC shared the same views. Jones, although not a Trinity graduate, was very close to the TC and especially Whewell. However, this Caius graduate initially confessed to Whewell that he found *metaphysics* and *metaphysical* to be utterly meaningless mongrel terms, which "offer themselves to every blockhead who will use them to fly in the face of anyone who puzzles him besides standing sentry to drive away the conviction of his ignorance whenever it approaches him. I want to shut them up in their own kennel." He warned Whewell, rather like Herschel did later, "As soon as metaphysical is allowed to mean deep & to be extended to all subjects then come darkness & poetical analogies made into abstractions & big words set up for things & abomination of all sorts." At this stage, Jones remained skeptical of Whewell's metaphysics and the divine power of ontology.[126]

Whewell's metaphysical notion of fundamental ideas clearly drew on Kant's philosophy of "pure intuitions" and concepts. Here the latter combined with the former via ideas produced by the "transcendental imagination"; the legacy of German idealism ran deep throughout Whewell, Hare, and Thirlwall. It was the tool to refute both pure empiricist and rationalist viewpoints. However, where Kant excluded established religion as the source of this process, the TC, like Coleridge, put it at the heart. Where Kant argued that the right morality led to the right religion and an ethical commonwealth, Whewell took the traditional view that the opposite was the case. In this way he and the TC were far more in tune with Coleridge's use of Kant. Nonetheless, the German philosopher played a key role in showing how empirical impressions were nothing without the mind to organize them. In an uncharacteristically precise fashion, Coleridge wrote in his *Table Talk* of 1832, "The pith of my system is to make the Senses out of the Mind—not the mind from the Senses." Interpretation of senses depended on the religious mind. This perfectly summed up the TC's disdain for the human-centric emphasis on sensory understanding.[127]

Whewell tried to defend his use of metaphysics in the philosophy of mind, mathematics, and language in his next letter to Jones. He also assured him there could not be any metaphysics of political economy, "for there are no peculiar principles of observation or deduction in that science" to bind it together as true knowledge. He reassured his friend, "You and [Thomas] Malthus belong not to the *metaphysical* but to the *ethical* school of Political Economy."[128] He later pleaded to Jones not "to tell our friends that I am bedevilled with German philosophy" and "a priori metaphysics." Whewell was rightly worried such an accusation would be adverse; you "will not be able to remove when I have convinced you . . . that everything which I believe is most true, philosophical & inductive."[129] His defensiveness was probably born of conservative Anglican suspicion of German philosophy that, as we have seen, played a prominent role in the reform movement in Germany.

For many in social authority, German metaphysics was dangerous and just as much a threat to established religion as the sensory philosophers (everything based on experience) or pure rationalists (independent of experience—such as algebraic analysis). Indeed, many of Kant's German contemporaries, such as Friedrich Heinrich Jacobi, saw Kantian philosophy as equally undermining social authority and higher education—just as the rationalism of the French Revolution had.[130]

In later life Jones would become much more supportive, if not a convert, of Whewell's metaphysical stance. In practice Jones's work was much closer to the approach, for example, of Kant's onetime student, Johann Gottfried Herder, and the perspective that history should view the past as a series of distinct ways of life.[131] In many ways Whewell's philosophy came to complement the *later work* of Schelling. The latter was Bunsen's favorite theorist, particularly after Schelling had been hired in 1841 to replace the godless Georg Wilhelm Friedrich Hegel in Berlin and purify the anti-Christian, republican, and revolutionary sentiments running roughshod through the city's streets.[132] Both men worked to reconcile science with religion. However, Terry Pinkard argues, those who went to all Schelling's lectures that year were left believing "they were hearing little more than a kind of oddly patched together apologia for contemporary state-sanctioned Christianity and for the authority of the monarchy." Whewell would have approved.[133] In many ways both men were attempting, as Damon Linker writes regarding Schelling, to produce a way to "combine science and poetry and, in turn, reintroduce a divine dimension to the disenchanted world bequeathed to us by the Enlightenment."[134]

For the TC one clear embodiment of this godless disenchanted world was the arch-utilitarian Babbage. By 1830 Whewell told Jones he thought "Babbage is come & gone wilder than ever."[135] Moreover, the only reason Whewell accepted a seat on the Royal Society in 1831 was, he claimed, "my dislike to be coupled with Babbagian sect of spewers or railers."[136] Whewell launched an attack on algebraic mathematics in his 1833 Bridgewater Treatise, *Astronomy and General Physics Considered with Reference to Natural Theology*. Here he argued French mathematicians like Lagrange and Laplace did not have the mind of a "discoverer, who, for the first time, seizes the principle which connects phenomena before unexplained, and thus adds another original truth to our knowledge of the universe." All these mere mathematicians did was describe the known law of universal gravitation: "But when [the Trinity man] Newton conceived and established the law itself, he added to our knowledge something which was not contained in any truth previously known, nor deducible from it by any course of mere reasoning." The process of induction was far more important than meager mathematical deduction. No matter how brilliant the mathematician, they paled into insignificance compared to the lone, pious, inductive genius—a person in touch with inner divine ideas that made sense of such observations.[137] The message of his Bridgewater Treatise was to underline that it was the task of natural philosophy to illuminate natural theology and not the latter to conform to the former.[138]

Indeed, when a curriculum concentrated on "the habitual and exclusive prosecution" of pure mathematics, Whewell warned, it can have "an unfavourable effect on the mind of the student." It makes them less receptive to "accept truths different from those with which his reasonings are concerned." Surely, with the prevailing Lucasian Professor of Mathematics Babbage in mind, he then concluded, "We conceive, for example, that a person labours under gross error, who believes the phenomena of the world to be altogether produced by mechanical causes, and who excludes from his view all reference to an intelligent First Cause and Governor." All deductive reasoners—pure mathematicians or logicians—were doing was developing and extracting known truths and presenting them as conclusions. The whole process of discovery was ignored: "these methods can generate no new truth." Indeed, all they did was "make men insensible to moral evidence and to poetical beauties."[139] No wonder a clearly outraged Babbage rushed to respond to such a smear in his unofficial *Ninth Bridgewater Treatise* of 1837. The following year he resigned from his Lucasian Chair.

Whewell could barely disguise his fury at Babbage's impish and unofficial response, which Hare referred to as a "hateful book" and in which the author confessed to being induced by Whewell's *Treatise* to publish his alternative view.[140] Indeed, Whewell's treatise had been deliberately inflammatory toward the view held by Babbage and his ilk. Consider his blistering attack on "deductive habits" in his chapter 6.[141] He wrote, with folk like Babbage in mind, "However rare the mathematical talent, in its highest excellence, may be, it is far more common, if we are to judge from the history of science, than the genius which divines the general laws of nature. We have several good mathematicians in every age; we have few great discoveries in the whole history of our species." Indeed, he concluded, "a mere mathematician or logician" pursuing a deductive, rather than inductive approach, leads to error—as opposed to "one who studies the facts of the natural world and detects their laws."[142]

Whewell's temperature always rose in public spats like this. In an open letter published to Babbage in the *Athenaeum* that same year (1837), Whewell went to the heart of his and the TC's primary fear: "I have endeavoured to explain, by pointing out that certain habits of mind may lead men to substitute for the deity, certain axioms and first principles as the cause of all;—to thrust some mechanic cause into the place of God, instead of raising their views, as great scientific discoverers have done, to some higher cause, some source of all forces, laws, and principles."[143] Babbage would have simply smirked and rejoiced at the reaction.

Everyone knew what was at stake. Babbage thought Whewell offensive for denying that pure mathematical algebraic analysis was incapable of finding new laws of the universe. As far as Babbage was concerned it was the most powerful tool for discovery and rested purely on human reason. On the contrary, argued Whewell, "certain algebraical formulae, and the knowledge when and how to apply them, constitute *the last step* of the physical science to which we can attain" (my italics).[144] In his second edition Babbage completely inverted Whewell's philosophy and wrote in his preface, "The first class of truths then (those of pure mathematics) appears to rest on necessity. The second (the laws of nature), on necessity and our external senses and internal consciousness [not divine fundamental ideas but a correctly functioning mechanistic brain]. The last (those of revelation), on human testimony [therefore floored and untrustworthy]."[145] Babbage was odious to the TC for having such total faith in human reason and denying an omnipresent deity and the direct word of God, not human testimony, via the apostles.

THE CAMBRIDGE CAMDEN SOCIETY (1839–1859)

Many of the emerging views among the TC manifested in the establishment of the Cambridge Camden Society in 1839. The society was created by three Trinity students, namely the Sherbourne-taught John Mason Neale, the etymologist Edward Jacob Boyce, and Harrow-educated Alexander Beresford Hope. The central aim of the society was to promote Gothic architecture, which they believed could revive the piety, spirituality, and religious perfection of the Middle Ages. This in turn could be used to inform Church reform. The society was in a sense Cambridge's answer to the Oxford Movement and gained a great deal of support, or perhaps more precisely policing, from Whewell and especially Thorp, who became the society's first president for the next twenty years. Along with Rose and another Trinity don, William Hodge Mill, Thorp was one of the fiercest High Churchmen at Cambridge.[146]

The Camdenians were, however, very different from the Tractarians. Owen Chadwick writes, "Newman and [Edward] Pusey began with the doctrine of authority and asked how best to execute its ordinances. Neale asked a different question. How shall men be led to worship? The Tractarians were concerned for truth and then for the issue in worship. The Camdenians were concerned for decoration, ritual, the structure and seating of churches, because these affect the way in which we worship." The model was Gothic, and the rituals adopted were those that had survived the Reformation and not Catholic doctrines.[147] For the so-called Tractarians, and a sympathetic Rose, ultimate authority came from the Holy Apostles and not the state; both groups, however, were fierce opponents of a mechanistic human reason.[148]

The strong "Romantic," poetical element within the Camden Society, taught to its members by the TC, led them to idealize the Gothic high-medieval period. This era was interpreted as intuitive, natural, and spiritual compared to the cold utilitarian era in which they were living. Gothic architecture was being promoted to provide Britain with a national and imperial virtuous representation. This was accompanied by a transformation in antiquarian, linguistic, and historical scholarship.[149] Both Whewell and Thorp shared this view but were also keen to ensure the society's members did not forget they were students and subservient to their Cambridge teachers.

A key remit of the society was to visit and study Gothic churches. It was believed that every detail of the building had a precise spiritual meaning. Whewell's inductive study, *Architectural Notes on German Churches: With Remarks on the Origin of Gothic Architecture* (1830), was popular reading among the society's members; the Teutonic (German, Scandinavian, and Anglo-Saxon) impregnation within the British past ran deep.[150] The society changed its name in 1845 to the Ecclesiological Society, by which time it was dominated by Beresford-Hope.[151]

Although Coleridge had refrained from historical study, his work pointed and inspired many to find purity in the Gothic medieval period. This combined well with Wordsworth's adoration of nature and the peasantry. This earlier period remained cleansed of the new commercial system that had ushered into the world a grimy industrial society, commercial selfishness, and false "rational" philosophy. Gothic architecture offered a virtuous and divine representation of something free of modern contamination.[152]

Whewell's student, Digby, who went on an architectural tour, idealized the chivalry of medieval life in his earlier book *The Broad Stone of Honour, Or Rules for the Gentlemen of England* in 1822. The volume was employed by many in the intellectual war against utilitarianism. Feelings from the heart usurped the prevailing obsession with rationality and was embodied in medieval chivalry. Hare's nephew, the Harrow and University College, London–taught Augustus J. C. Hare, much later described the work as "that noble manual for gentlemen, that volume which, had I a son, I would place in his hands, charging him, though such prompting words would be needless, to love it next to his Bible."[153] Medievalists, be they literary writers or historians, were drawing on studies of peoples to reveal the superiority of Britain's organic, liberal, and faultless institutions. This was perhaps most revealed in novels such as *Ivanhoe* (1820) by Walter Scott, which celebrated one of the last fictionalized noble Saxon families in twelfth-century England. Britain's institutions stemmed from the purity of this period.[154] By the second quarter of the nineteenth century the TC looked to the medieval world of Christian chivalry as a virtuous and pious period.[155] The Oxford Movement proper (Tractarians) preferred plain churches and plain worship, much prayer but relatively little music, whereas "ceremonial doctrines" were mainly associated with the Cambridge Camdenians, the true begetters of the so-called second Oxford Movement from the 1850s.[156]

It would be Coleridge's *Aids to Reflection* (1825) that provided the most penetrating critique of evidence writing such as Paley, Unitarianism, and rational theology. As a student at Jesus College, Coleridge had at the time been a Unitarian and had supported Frend's attempt to allow non-Anglicans to hold government positions. Coleridge now, however, was a fierce opponent of the creed. He described Unitarianism as the religion for a man "whose Reason would make him an Atheist but whose Heart and Common sense will not permit him to be so."[157] Coleridge, like the TC, was not saying faith conflicted with reason but rather that it was the apex of reason. In his later work, *Constitution of the Church and State* (1830), he stressed the "nationality" of the Church as established by the state and its key role in education.[158] His work is important in understanding the perspective especially of Whewell, Hare, and Thirlwall.

These initial Trinity members of the Cambridge Camden Society would have also enjoyed the teaching of Regis Professor of Greek Monk.[159] Prior to Christopher Wordsworth's appointment as master in 1820, he had been the favorite for the post.[160] His views of a university education were underlined in a sermon in 1822: "An institution whose professed object is the support of true religion by educating able persons for Christ's ministry." The university's role was teaching established knowledge and not producing new knowledge.[161] Whewell successfully lobbied Monk to make Hare, also like Monk a former Charterhouse and Trinity student, to be his replacement in 1824.[162] Later in 1841, on becoming the new master of Trinity, Whewell diplomatically wrote to Monk thanking him for his mentoring, "I look back with great pleasure to the time when you selected me as your Fellow-labourer in the tuition, and consider that as the turning-point of my life, which decided my course to be what it has since been."[163]

THE EXTENDED TRINITY CIRCLE

Other prominent Trinity students and followers of the TC included the Etonian, politician, and poet Winthrop Mackworth Praed, who toured Germany with Hare in the summer of 1828; the Bury St. Edmunds (King Edward VI)–educated poet and writer Edward Fitzgerald; the Etonian, poet, Cambridge Apostle, and son of the historian of the same name Arthur Henry Hallam; the historian and politician Thomas Babington Macaulay; and the St. Paul's–educated Cambridge Apostle Robert Leslie Ellis, who jointly edited the works of Francis Bacon in collaboration with two other Trinity graduates, namely Douglas Denon Heath and James Spedding. Additional Trinity students included the professor of Greek at University College, London from 1831 until 1876, Henry Malden; the poet, Cambridge Apostle, and politician Richard Monckton Milnes, 1st Baron Houghton; the Cambridge Apostle Joseph Williams Blakesley; the Etonian, poet, and hymn writer John Moultrie; and the writer and politician Edward Bulwer-Lytton. Although based at St. John's, the third child of Coleridge, Derwent Coleridge, was also a firm favorite of Hare and Thirlwall—as was the apostle and Caius student William Bodham Donne.[164]

This star cast of predominantly Trinity men were primarily disciples of Whewell, Hare, and Thirlwall during the crisis years of the second quarter of the century. Milnes later reflected that Thirlwall was the most impressive man he had ever met and soon became one of Thirlwall's closest friends in London. The only other house Thirlwall visited while he was in London was that of "Mrs William Bayne, daughter of Professor [George] Pryme of Trinity and widow of a Trinity Fellow, whose son, sponsored by the Bishop, was named William Thirlwall Bayne."[165] Pryme had also attended Sedbergh, Sedgwick's school, and played cricket with him. They later "became very intimate friends" at Trinity.[166]

It is worth noting that as well as being Trinity students Spedding, Kemble, Donne, and Fitzgerald had all gone to Bury St. Edmunds Grammar School (King Edward VI). Here they were taught by the former Trinity student, the headmaster of the school, and William Blake scholar Benjamin Heath Malkin. Malkin was responsible, it seems, for the making of many early Trinity apostles before leaving the school to become, interestingly, the first professor of history at the new London University.[167] The school's connection went back much further. For example, another Trinity fellow, Robert Edward Garnham, had attended the institution during his father's period as master. Interestingly, Garnham resigned from the Church of England in 1789 after reaching the conclusion that the Scriptures were both the revealed word of God but also human testimony.[168]

Coleridge's emphasis on an intuitive knowledge of God, nurtured by the right morality and feelings (the outward sign of God in man), directly pervaded the outlook of the TC and its extended movement. This, in fact, was not far from the thinking of the eighteenth-century French philosopher Jean-Jacques Rousseau, with his emphasis on ethics as the mainspring of education stemming from one's conscience, an innate-divine instinct that separated man from animals. Such morality, rather than human reason, was the true reflection of God's world.[169] It was thanks to Coleridge that Hare, for example, owed "more than to any other man" the ability "to trace the path which connects human learning with divine, the faith on one age with the faith of another, the sense of man's grandeur with the sense of his pettiness and sinfulness."[170] He bemoaned the "boast of the age" and poured scorn on the claim to supremacy over the past—that "our ancestors were utterly ignorant and blind." By contrast we were connected to the past and could not, must not, sever this sacred link. To do so would mean we all "must fall to the ground."[171]

Coleridge and the TC provided a philosophy of knowledge as divinely inspired. The human will and soul, not design in nature, was proof of God's existence; proof of design came from within. An appropriate education that reflected this was therefore crucial to ensuring that Cambridge's undergraduates, the nation's future spiritual and political leaders, did not divert from the divine road; it was in the process of recognizing one's own spiritual nature that one knew God. This was the unseen world and central to nurturing genuine morality. The world had become a machine with no soul. The task of putting the soul back in the machine became the quest of the TC.

CHAPTER TWO

PUTTING THE SOUL BACK IN THE MACHINE

> Nobody can deprive us of the Church if they would, for it has the affections of the people in its favour; and, even if it had no support from Government, would be supported by the conviction of its truth and of the good it does to all. . . . And I have a deep conviction that Providence, which has guided it through so many dangers, and made it so great a blessing to the nation, will not allow any misfortune to befall it which will permanently interfere with its usefulness as the bulwark of religion and good order.
>
> WHEWELL, MAY 8, 1835

It was hoped, by a minority section of the Anglican clergy during the early nineteenth century, that by importing German literature and, by a larger section, furthering the works of Coleridge and Wordsworth, the intellectual and spiritual activity of England could be revived. What German philosophy provided was an organic-spiritual structure. This contrasted with the mechanistic-material foundation underpinning British rational reformers. The theological struggle sweeping through the country and the old universities was evident everywhere—even in the physical geography of Cambridge.

During the early 1810s the city was embroiled in a fundamental debate concerning the distribution of the Bible: "The walls of the town were placarded with broadsides, and posters were sent from house to house."[1] Should the Bible be distributed with or without the Book of Common Prayer? Herbert Marsh, the staunch anti-evangelical Anglican conformist, cousin of the expelled Frend, and Margaret Professor of Divinity, had no doubt: the prayer book must be distributed as a companion to the Bible to prevent false interpretations. The King's School– and St. John's–educated don pleaded, "It is not the Bible itself but the perversion of it, the wresting of the scriptures . . . by the 'unlearned and unstable,' with which England now swarms, whence the danger proceeds." In short, the poor had to be taught the Bible by the Established Church if the security of Church and state was to be preserved.[2]

However, a growing sentiment was arguing for the freeing of the Bible from Church authority. The heart of this debate centered on social authority and the role of the Anglican universities. For Marsh, this could only be entrusted to the clergy of the Established Church, for which Cambridge and Oxford were the principal source of supply. He was also a close adviser to Lord Liverpool over economic matters during the Napoleonic Wars and was rewarded in 1816 with the see of Llandaff,[3] where he succeeded a Trinity man of another era, Richard Watson. Marsh was also the great friend and obituarist of the Unitarian-minded tutor Thomas Jones. Jones was a senior wrangler and First Smith's Prize man (1779) who became head tutor at Trinity and mentor of Sedgwick.[4]

THE BRITISH AND FOREIGN BIBLE SOCIETY

The British and Foreign Bible Society (est. 1804), the source of the view that the Bible should be distributed on its own, was composed of evangelical Dissenters as well as evangelical Anglicans. Its spiritual home was Clapham in South London.[5] Marsh fervently believed that many who supported the Bible Society did so simply because it challenged the established socio-structural hierarchy. As he made clear: "If [the Bible Society] is attended with evil, it receives a tenfold augmentation, by being fixed in a seat of education." Setting up a branch at Cambridge had serious repercussions: "What must it be when established in a place, where the youth of this kingdom *will be taught to embrace it*, and to *disseminate* that evil throughout the British dominions." Marsh's main opponent, the formidable evangelical Lucasian Professor of Mathematics and master of Queen's Isaac Milner, accused him of being paranoid and blocking the good that would obviously proceed from the distribution of the King James Bible by itself.[6] He was joined by Simeon, the Jacksonian Professor William Farish, and Rose's former tutor, the professor of minerology Edward Daniel Clarke, in setting up and promoting the society.[7] Milner had tried to become master of Trinity in 1789 and 1798 because he considered it the last remaining sanctuary of Jacobinism within the university.[8]

By contrast the then fellow of Trinity and younger brother of William Wordsworth, Christopher Wordsworth, agreed with Marsh and published his *Reasons for Declining to Become a Subscriber to the British and Foreign Bible Society* in 1810. Marsh and Wordsworth were also deeply concerned about the spread of Lancaster's nondenominational education system for the poor, which included a school set up in Cambridge in

1808. The next year, in a joint venture with Wordsworth's friend, the High Churchman and fellow Hackney Phalanx member Joshua Watson, they established a rival national society to provide an education that embodied the principles of the Church of England—the National Society for the Education of the Poor in the Principles of the Established Church—as a direct competitor to the Institution for Promoting the British System for the Education of the Labouring and Manufacturing Classes of Society of Every Religious Persuasion.[9]

The threat of the Bible Society toward the Established Church was also something that came to be shared by the TC. Consequently the position of the Anglican Church was in danger from abstract rationalists, sensory empiricists, growing irreligion and sectarianism in industrial regions, as well as evangelical enthusiasts both next door and within.[10] The theological doctrine of the Church of England was being dangerously eroded and needed rescuing.

When Babbage announced the creation of the Analytical Society, he had done so partly as a parody of the Bible Society: "[We] drew up the sketch of a society to be instituted for translating the small work of Lacroix on the Differential and Integral Calculus. It proposed that the Society should have periodical meetings for the propagation of D's; and consigned to perdition all who supported the heresy of dots. It maintained that the work of Lacroix, like the King James Bible, was so perfect that any comment was unnecessary."[11] Babbage's brilliant caricature of Christian evangelical faith and the Bible Society was not just humorous, however, but had ramifications that were potentially as subversive; despite their initial enthusiasm, this was something the TC soon realized.

To recap, Babbage and Herschel believed the foundation of knowledge and progress was human reason empty of an omnipresent God. The axioms underlying the creation and subsequent functioning of the world were the same as the operations involved in logical human thinking (an algebraic analytical machine) and consequently human invention. Therefore the plan of the world and proof of God's design could be found in the operations of human intelligence; observations needed to go through the mind's mill before being correctly processed and classified into types. This was deism and in step with a certain Dissenter ideology and utilitarian confidence in man's own abilities free of the Church and divine intervention. The idea of nurturing divine faith based on revealed knowledge was dismissed. For men of Babbage's ilk the only thing revealed was the myth of Adam and the Fall. The prolific transnational political radical Thomas Paine had slightly earlier described it as "a fraud of the Christian system to call the sciences *human invention*; it is only the application of them that is human. Every science has for its basis a system of principles as fixed and unalterable as those by which the universe is regulated and governed. Man cannot make principles, he can only discover them."[12] By contrast most Anglican Christians believed man had fallen and his reason now led us astray, and consequently only a regenerated or sanctified reason (combined with the other human faculties) could gain an image of the world as it really is. Salvation and a true understanding of the creation of knowledge were needed more than ever.[13]

Babbage probably cultivated some of his republican and blasphemous outlook during his early teens when he visited Lucien Bonaparte, the exiled younger brother of Napoleon, at his home in Worcestershire. Herschel's religious dissenting views were later to cause anxiety for his wife, Margaret Brodie Stuart: "We must have our Herschel a decided and recognised Christian, a 'city set on a Hill'—but perhaps it must be attempted more on our knees and in our conduct and conversations than by our arguments . . . he confesses, that he was not accustomed to work the life of a humble and sincere Christian living and rejoicing in their God."[14]

For Herschel and Babbage, an emphasis on tradition and faith in the word of Scripture was not necessary or indeed any use in the manufacture of objective knowledge. The process of discovery was a product of human reason described by the logic of algebraic analysis. It was not dependent on emancipating some deep divine inner idea now overladen by human corruption after the Fall. The key was to ground and organize the operations of intelligence into an efficient, atemporal system of production. Gathering the observations via induction was crucial, but their meaning was absent until they had been pushed through the mechanical operations of the analytical mind.[15] Whewell was horrified and reflected in his diary (1825), "The view entertained by those who do not see the necessity for a perpetually sustaining and preserving Providence in the world, seems to be this, that the material notions and elements of the universe may have certain *laws* impressed upon them—or certain properties in their *nature*" that allow them to combine, operate, and move without the need for divine intervention. Under this view there was no need to consider "an immaterial and intelligent agent," and as such, "we have brought the idea of *the world as a machine* much more distinctly into our view" (my italics).[16]

The attraction of this view was very clear, namely its simplicity and ease in making generalizations. Once a law, say of motion and force, was obtained, it could be applied to anything that moved. However, Whewell wrote, this is to miss the point: "We have discovered no new power, no new agency in nature; we have only found the rule according to which the results are produced." The actual origin of these rules is the

Creator: "We can read it [the Book of Nature] so far as our minds are of the same nature with his." These "laws of nature . . . are laws according to which the creator causes matter to act and work." Everything for Whewell, as for William Wordsworth, was a result of an active divine legislator—from the leaf that dropped from a tree to the bubble that burst in a stream. It was "strange and insane," he sighed, that certain men should fix their attention solely on these rules and therefore "deify as it were the maxims of his government when we deny that he governs it all." It was not independent self-acting laws of motion impressed into matter but, on the contrary, living divine ideas actively embodied in matter: "Every new law that we discover does thus, in proportion to its generality, bring us nearer and nearer knowledge of the thoughts of God as they regard the material world." By contrast, such an approach was far from applicable to the moral world since the laws discovered were of a much more limited generality and far less accessible—certainly by human reason alone.[17]

Nonetheless, there was proof of intelligent design in the moral world as opposed to simply Paley's emphasis on the material world. As Whewell put it, also in 1825, man was perfectly adapted to society thanks to his "constitution of affections" that "unfolded by habit and by social intercourse." These affections tended to work together for good, happiness, and the forging of an elevated character in man—this thus demonstrated "a system of design of which his creator is the author." You could not cultivate the intellectual without the appropriate religious feeling and accompanying morality; to do so would lead to darkness: "The Christian principle of action is not a selfish one—it is a principle of love. That it is not a mere hope of reward, a bare calculation of the result, but it is a constitution of the affections and a tendency of the heart." This feeling should dominate your life, Whewell urged: "If therefore the love of God, his salvation, of his word, of his righteousness do not find a place in the heart it will be followed by the love of the world, of pleasure, of vanity and of sin."[18]

However, selfishness, vanity, and love of material goods was usurping this traditional religious view. Such a momentum had commenced at least since the early eighteenth century and captured in Bernard Mandeville's *Fable of the Bees: or, Private Vices, Publick Benefits* (1714). Instead, such values were now worshipped and legitimated by a philosophy of utilitarianism. This in turn was verified by the doctrines of Ricardian abstract political economy. There was no room for God in this human-centric world. Happiness derived via material wealth and not spiritual purity. The world had turned upside down, creating anxiety among men like the TC.

ORGANIC CAMBRIDGE AND THE SPIRITUAL WORLD

The growing utilitarian-materialist outlook clearly went against a powerful prevailing view that society had its own temporal inner life of growth and organic change, which could ultimately be known only by the Creator. However, and it was a crucial however, the TC believed that through a lamp shining on true faith, a mirror of this was possible with the light becoming clearer over time. It was vital to stay on the true path.[19] Ideas slowly emerged from knowledge via a spiritual-dialectical interaction of the mind and the material world. Institutions formed through centuries of experience could not be reduced to first principles; the British Constitution had no theoretical foundations or formulae and was certainly not erected on such faulty abstract rational propositions as France had implemented since the country's Revolution. Society was a spiritual bond among those who are dead, alive, and yet to be born; the British Constitution and its institutions were an organic, ongoing creation born of a shared understanding.[20] Spearheading a distinct, evolving, organic Constitution with a National Church at its heart was Coleridge, especially as described in his *On the Constitution of the Church and State* (1829). Like the Constitution, the mind was also in a state of growth.[21] This needed careful nurturing and protection from the corruption of an artificial intelligence.

However, traditional social order, sustained by the sanctions of divine authority and historical precedence, were being challenged by attacks on established religion and, intellectually, by efforts to formalize and make visible the mechanistic operations of intelligence. Human creativity, this subversive view held, was not bestowed by the grace of God; rather, it was a product of thinking correctly. Imagination, inspiration, and even having a soul were not part of reasoning and should be expunged.[22] Morality derived from religion was redundant and being replaced by mechanical calculation—with the objective of gaining the greatest happiness for the greatest number of people. Human behavior and moral conduct were nothing more than a prescriptive association of sense impressions and laws underpinning self-acting matter. This was the perspective, most prominently in Britain, of Bentham and his friend James Stuart Mill, the physiological materialism of William Lawrence, and a theory Babbage subsequently tried to build into a machine and others utilized to change society. No institution, including Cambridge, was immune from the related momentum for reform that for many ultimately included severing religion from a university education.[23]

An early sign of Whewell's growing conservatism and hostility can be found in his first pedagogical text on mechanics in 1819, which was carefully constructed to repel potential French analytical subversion:

> It appears as if their [French mathematicians'] attachment to the forms and processes of pure analysis, which they have cultivated with such signal success, had given them a disrelish for the more physical and inductive part of reasoning, and made them completely indifferent as to the manner in which they arrive at that part of the subject where the machinery of analysis begins to work. Hence those principles which mechanics must borrow of experiment are often made to depend on abstract reasoning's and artificial definitions: or introduced as self-evident, with some slight notice as to their agreement with matter of fact.[24]

In other words, French algebraic analysis did not reveal the process of discovery but insisted on making the path *conform* to its process of applied human abstract logic, at least defining a discovery found by other means. Human logic and algebraic analysis were forming a dangerous alliance. Whewell knew he had to introduce such Continental algebraic analysis into the Cambridge Tripos; the momentum was too strong, but he cleverly did so fully dressed in the traditional university clothing of Newtonian mechanics (a proven thing) founded on proper geometrical discovery. It could be used to describe such a specific astronomical cosmology but not as tool of discovery. Herschel could barely restrain his disappointment, claiming that Whewell's work would have "been much more widely circulated, had you conformed a little more to the taste of the age and a little less to the University."[25] Acknowledging the dead end at Cambridge, he now fled to London to further his analytical reforms with Babbage, beginning with the establishment of the London Astronomical Society in 1819 and involvement with more sympathetic metropolitan circles.[26]

By contrast, Whewell stayed planted in Trinity and began to synthesize his intellectual, theological, moral, and political views. The machine philosophy of Babbage and Herschel would not reveal the "vast and solid edifices of truth" that exist; there were divine fundamental ideas that made sense of things such as "space," "force," "time," "number," "fluid," "grammar," and architectural "verticality." A purely empirical and mechanistic origin to truth was superficial, failing to reach a "far deeper root in our minds."[27]

While Babbage and Herschel were busy trying to reform science from a base in the center of the capital city, Whewell went to visit Wordsworth in the rural majesty of the Lake District with Sedgwick in 1821 and again in 1824.[28] For Whewell divine "fundamental ideas" were the building blocks of science. As he later wrote in his *History of Scientific Ideas*, "Without our ideas, our sensations could have no connexion; without external impressions, our ideas would have no reality; and thus both ingredients of our knowledge must exist."[29] This was the perfect answer for assimilating science and experimentation, while retaining the overall presence of God. These Trinity men were on a quest to preserve the status quo and not to change the world.

Thirlwall moaned to Bunsen, "The predominant inclination just now in England more than elsewhere, is to regard things from the side of utility, or rather of gain. This commercial spirit values even art and knowledge simply as wares which shall only be brought to market in order to make the best use of them; not imparting but exchanging shall be the aim of the acquired knowledge and skill." He concluded this was due to the shattering of tradition by "the perversities and extravagances of powerful minds which have broken loose from the old arbitrary restraints, and are governed by no self-acknowledged, well-grounded principles."[30] In January 1822, revealing his reading of Coleridge and Wordsworth, Whewell wrote to his sister, Ann: "I can never believe that the imagination, the fancy, the taste of man, would have been so exquisitely constituted as they are, if it could be a duty not to exercise them. On the contrary, it is only when we consider man with all his faculties unfolded, that we can have any proper idea of his relation to his Creator and of the wonders of Divine Wisdom and Goodness."[31]

Creativity, as emphasized especially by Coleridge, was a product of the imagination, not formularized rules. Drawing on Kant, our sense experiences detect things not as they necessarily really are, but as appearances, and it is the imagination that gives them meaning, thereby creating the external world, transcending impressions to create a unified whole. This was the ideal argument to combat the growing mechanistic and sensory worldview.[32]

Only those fundamental ideas that had been established should be taught. As such, new subjects without the necessary foundation of fundamental ideas—such as political economy (the antithesis for the TC of Christian morality) and the new sciences like chemistry, meteorology, and geology—should not be part of the curriculum.[33] It was up to a Cambridge liberal education to create the necessary habits to develop the divine faculty bestowed by God of established fundamental ideas; God's mind could only be found in our soul. This should be the heart of a Trinity and Cambridge education. A successful student would be one who achieved this without being consciously aware of doing so; such divine faith should be tacit, just like riding a horse or walking. This was central to the TC.

THE "INMOST DEPTHS OF THE SOUL" VERSUS "CLERKS IN THE COUNTINGHOUSE OF THE SENSES"

Total faith in human reason, atemporal axioms, and the march of steam philosophy severed the link with the past, attempted to build institutions afresh with no precedent, and thus destroyed the sacred living body of human relations. Fallen France and the growing overcrowded, immoral, dirty industrial regions of Britain were all proof of this. When Babbage wrote, "I wish to God these calculations had been executed by steam," the TC would have despaired—that faith in a mechanical process was far more credible and trustworthy than divine ideas. The ideal education was a union of established science and the Established Church—the two met in man's relationship with the Creator and not "to books and to formulas." It was through these means that our faculties and morality were revealed.[34] The romanticized English shire was being destroyed by the growing industrial, urban, and political re-mapping of a purely human-centric creation devoid of God. Whewell preached, "Morality, to be real, must not be merely a set of doctrines, but an influence upon men's hearts and characters."[35] The soul had to be put back in the machine; the cogwheels replaced by roots, and the dead metal structure substituted by a living plant.

The British poets William Wordsworth and especially Coleridge had, at least for the TC, revitalized Christianity, with the latter emphasizing intuitive knowledge of the unseen world and the role of the imagination. Whereas the new philosophy sought to discipline the imagination and, ideally, expel poetry, for these men it was the opposite—a glimmer of truth revealed by the lamp of a free imagination.[36] This was only possible by cultivating true belief and reaching as pure a spiritual state as possible. The realization of the latter could be seen in poetry. As Coleridge earlier wrote in his *Biographia Literaria* (1817), "The poet, described in *ideal* perfection, brings the whole soul of man into activity. . . . He diffuses a tone and spirit of unity, that blends, and (as it were) *fuses*, each into each, by that synthetic and magical power, to which we have exclusively appropriated the name imagination. This power . . . reveals itself in the balance or reconciliation of opposite or discordant qualities."[37]

Also in 1817 Whewell reflected in his diary on the eternal struggle, throughout history and countries, of reason and imagination. The latter was the name "which may be made to satisfy all that part of the feelings of our minds which is not analysable by or into reason." It was these feelings that knitted us together with nature, "which makes us feel the influences of mountains and skies of the wind and the sea," all of which lifted us out of the everyday world to a higher nature. Pure human reason was incapable of seeing any of this; indeed, "this mysterious principle" vanishes when touched by analytical reason. For Whewell poetry was the spirit of man and worked on his soul.[38] In 1820 he again noted in his diary how much he enjoyed Wordsworth's *The Excursion: Being a Portion of The Recluse—a Poem*, first published in 1814. He especially liked the Solitary's account, "which is beautiful," with Wordsworth's "favourite theme" being "the inadequacy of the understanding alone to reach truth."[39] True understanding required a divinely induced imagination.

Imagination was the inspired impetus to poetry and a mirror to the divine creative process underpinning the universe. Thus there were no rules outside of us in this development—to claim such was to simply reduce it to "a mechanical art."[40] As Coleridge earlier wrote, in his discussion of William Shakespeare as a genius, "By evolving the germ within by the imaginative power according to an idea—for as the power of seeing is to light, so is an idea in the mind to a law in nature." Shakespeare was the ultimate example of the preeminence of the mind over external nature.[41] Coleridge was clearly drawing on German philosophy that emphasized an organic chronicle—an ongoing history growing from the soil of the preceding period.[42] This informed the TC, in varying degrees, especially Hare, Thirlwall, and Whewell. It also primed Sedgwick's view of education and Jones's notion of an ethical economy.

Within the Trinity context, induction, poetry, philology, history, and philosophical idealism were harnessed to help combat the rise of a materialist reason and abstract political economy spearheaded by men like Bentham and Ricardo.[43] Possibly, the earliest public attack on the utilitarian world by the TC was made by Rose in 1818, when he chastised Bentham for his abusive critique of the Anglican doctrine in his *Church-of-Englandism and Its Catechism Examined*. In this work Bentham condemned the Church for using its catechism rather than the Bible; that this was not the religion taught by Jesus Christ.[44] Bentham was, Rose argued, ignorant of the Scriptures and the foundations of Anglican theology. Ultimately for Rose, such an intelligent man as Bentham had erred so badly because he reduced his faith to his own human strength of reasoning; he was the perfect example of corrupt human thinking after the Fall. The greatest threat to the Church, in this sense, was the dangerous use of human reason rather than any questioning of the Church's catechism.[45] This fear was something that pervaded the TC, although their solution was not, at this time at least, all the same. Rose became even surer that the Church should have a narrow but uniform doctrine and be on guard against all diversity. By contrast, at this point Hare and Thirlwall were advocates of a broader, more encompassing doctrine, with

Whewell somewhere in the middle, although ultimately veering much more to Rose's stance by the late 1820s.

The TC's adverse views toward these doctrines were exhibited, debated, and refined during Whewell's, Hare's, and Sedgwick's numerous visits during the 1820s to Hyde Hall, the home of the Tory Sir John Malcolm. Malcom was a self-made Scottish soldier in the East India Company before becoming the private secretary of Sir Arthur Wellesley (later Duke of Wellington) in 1801. He saw service throughout much of the Middle East and eventually became an MP in 1831, working with Wellesley to oppose the Great Reform Act of 1832. Hare wrote, "The house in which, above all others I have ever been an inmate of, the life and the spirit and the joy of conversation have been the most intense, is a house in which I hardly ever heard an evil word uttered against anyone." It is also true that the pious lads were enchanted by Lady Malcolm and their young daughter, Kate.[46]

Meanwhile Sedgwick, as a proctor in 1827, patrolled Cambridge checking the morality of students and no doubt associating poor behavior to changes in this ever-encroaching human-centric intellectual climate. Whewell told Lady Malcolm, "[he] walks about the streets with three tall men following him who are called 'bull-dogs,' and taking cigars out of the mouths of all the dissolute young men who use them."[47] A year later Whewell became professor of mineralogy and took the opportunity to escape Cambridge and travel to Germany, Switzerland, and Italy. Here he boned up on his mineralogical knowledge, German philosophy, and architectural understanding.[48] He was succeeded in 1832 by the St. John's graduate William Hallowes Miller, who flourished in the position.

All the TC were getting increasingly bad-tempered and taking out their frustrations wherever they could. The tension in Cambridge was getting worse. The university proctors and other university officers were assaulted by a group of men, resulting in some of the miscreants being caught. At the summer assizes eight were put on trial; one was cleared, and the rest were sentenced to prison for up to a year. Student resentment toward the proctors continued. In 1829 a large number gathered outside the Senate House and started hissing and throwing objects at the senior proctor and St. John's don Alexander M. Wale and his team as they left the building. Some of the students were caught and punished. However, the proctors felt more needed to be done. The result was a tightly worded letter from the vice-chancellor and heads of the colleges that if any student was again involved in abusing the proctors, they would be expelled.[49] Nonetheless, this led to the resignations of these men who policed the university.[50] As far as the TC was concerned, the new philosophy and impatience for reform was leading to disrespect and a physical challenge of everything—including the university.

Thirlwall much later warned a gathering at his old school, Charterhouse, to be on guard against these revolting, selfish, and sensory reformers: "Every long-established society subsists in and upon the monuments of the past, reared and transmitted by successive generations." This organic evolution was vital, for "without this inheritance the face of the globe would never cease to be a dreary wilderness. Hence there is at first sight the appearance of a contradiction between the testimony of our senses and language of Scripture." It was precisely through a "training of the whole man, unfolding all his faculties, and regulating their employment with a view to his supreme good," that the right platform for instilling the appropriate intelligence and morality into the country's future leaders took place.[51]

Hare preached to Cambridge students, in almost Dickensian fashion, of his increasing concern at the prevailing condition of England: "Our Thoughts are ever dwelling on outward things, and are mostly content to serve as *clerks in the countinghouse of the Senses*: and it is only with labour and difficulty, and after a long discipline and training, that we can fix them steadily on *the world within us*" (my italics). It was only by intensifying and preserving the close relationship between the university and the Established Church that true knowledge could be assured.[52] Instead, education seemed to be increasingly oriented to the manufacture of the type of character Charles Dickens satirized in his depiction of Thomas Gradgrind in *Hard Times* (1854).[53] Hare, according to his sister, Maria, was so angry that "I think he will be obliged to end by living in Germany, he is so much annoyed by the present system of things in this country—by the overpowering commercial spirit which fills everything."[54]

This shared notion of divine faith was intrinsic to Whewell's philosophy of science. The following from Hare could well be a discussion of Whewell's notion of fundamental ideas: "Faith is that power or faculty in man, which gives substance and reality to such things as are not objects of sight, and which fills him with a lively assurance of the things he hopes for." Without faith "the Intellect either worships itself, or dashes to atoms on the rocks of the Senses." This relentless condemnation of the poverty of human senses, self-love, and bold anthropocentric confidence to control the world was central to the TC's fear: "The objects of Faith are not the things which lie beyond the reach of Reason, but the things which lie beyond the reach of Sight, the things which are unseen, the things which as yet are objects of Hope, and which therefore must be remote from the Senses." Since all truth was ultimately divine, only those with true faith could access such truths. Hare warned, "When Faith dies away the heart of a nation rots. . . . All the faculties with which man was endowed in order that he might turn this world into the temple and garden of God, busy themselves in building and decking out an earthly Pandemonium." If you needed proof of this, he claimed, just look, once again, at what had happened to France during and since their Revolution: a false world built on the single faculty of human reason.[55] The TC truly believed that without such divine faith the same human-made disaster lay ahead for Britain. Whewell preached that the "Will of God" was the formation of our morality and this was being dangerously undermined. Humans, as instructed by Scripture, expressed faith through obedience to God.[56] The problem was finding a way, as we shall see in the next chapter, of providing the necessary building blocks for nurturing such belief.

CHAPTER THREE

GERMAN BIBLICAL CRITICISM AND GUESSING

> Rule 1. There is a falsity involved in all conclusions of science which contradict our conviction of the being of God. . . . This argument differs from the common argument for design. . . . Laws are not machines—are not first constructed and then left to act.
>
> Rule 2. There is a falsity involved in all conclusions of science which contradict our conviction of the goodness of God.
>
> Rule 3. There is a falsity involved in all scientific conclusions which contradict the revealed will of God.
>
> Rule 4. There is a falsity involved in all scientific conclusions which contradict the rules of duty.
>
> WHEWELL, FEBRUARY 18, 1827

As well as being a fervent opponent of the Bible Society, Herbert Marsh was the first to import German biblical criticism into Britain from Leipzig at the end of the eighteenth century, and was the first to teach theology in English at Cambridge. He was probably a major spark to the TC's focus on German literature, especially via his well-attended lectures. From 1793 to 1801 he translated the work of the Prussian biblical scholar Johann David Michaelis on the New Testament.[1] Marsh was especially keen to investigate the origin of the first three gospels and is referred to in Thirlwall's subsequent translation of Friedrich Daniel Schleiermacher's critical essay on the Gospel of St. Luke in 1825.[2] Terry Pinkard writes that Schleiermacher was especially dissatisfied, throughout his writings, "with German Protestant Christianity as being little more than a fragmented, lifeless bureaucracy far more interested in enforcing small details about doctrine than pursuing any kind of truth."[3]

Since 1810 Schleiermacher had held a theological chair at the recently established University of Berlin. He was a leading exponent of higher criticism and the scrutiny of texts, including the gospels. Thirlwall's translation was designed to dilute Anglican fears over the German approach and to introduce it to a wider public. It was, he claimed, now widely believed by the educated that the "sacred writers" of the Christian doctrine were not "merely passive organs or instruments of the Holy Spirit," but that they had expressed their own understanding of God's truth according to the customs of their own times and places. "This doctrine has been so long abandoned that it would now be a waste of time to attack it. When I say it has been abandoned, I mean of course only by the learned; for undoubtedly it is still a generally received notion." This was deliberately inflammatory, if not a little pompous and naive, at a time of evangelical upsurge. However, Thirlwall was no Schleiermacher. At this youthful moment he envisaged "a 'religious community' of free agents," who would abandon miraculous aspects of Scripture to create a far more robust, latitudinarian National Church.[4]

Both Thirlwall and Hare believed, following Niebuhr and Bunsen, that history correctly constructed would reveal the genuine divine doctrine; criticism, in this sense, aided faith. In his translation, Thirlwall demonstrated that he was familiar with other German critical scholars such as the theologian and orientalist Johann Gottfried Eichhorn and Heinrich Eberhard Gottlob Paulus, along with the theologian and close friend of Schleiermacher, Gottfried Christian Friedrich Lücke. The "new hermeneutics" of German philosophy was, as Nigel Leask writes, "a counter-enlightenment tradition rejecting the analytic, a-historical standpoint and asserting the 'polar' doctrine that 'like may know like.'"[5]

This approach caused a rift in the TC. Having initially been sympathetic and part of the undergraduate Trinity German reading group, Rose now related far more to the Oxford Tractarians of the following decade over Scripture. He was dismayed at Thirlwall's translation of Schleiermacher and his endorsement of German higher biblical critics, which he equated with "the most violent English Unitarians."[6] Rose disliked German biblical criticism as much as French algebraic analysis and utilitarianism. The Scriptures demanded total acceptance, and as such, the application of the single faculty of human rationality to critique them was simply a false and dangerous approach.[7]

Significantly during this period Thirlwall was enjoying the company of the biblical scholar, theologian, and friend of Schleiermacher, Wilhelm Martin Leberecht de Wette, and escorted his then-guru Schleiermacher on a visit to Cambridge from London.[8] Rose greatly disliked the works of de Wette and wrote, "I never saw any writings more offensive than the early ones of de Wette," and thought the work by Schleiermacher smacked of pantheism, that is, equating God with the material or forces of nature. This was also close to other German philosophers, such as Herder, Goethe, and Schelling.[9] Any friendship between Thirlwall and Rose had now gone.

Rose's first public warning about importing the German critical method came in a sermon, "The State of the Protestant Religion in Germany," given to the university in May 1825, which was clearly a response to Thirlwall's recently published work. Rose would have felt much happier with the German revival and Confessionalist movement, led by the anti-critical method biblical scholar Ernst Wilhelm Hengstenberg, who preached in 1825, "The Messianic idea in the Old Testament is no human invention, but truly from God. This idea is one and the same in all prophets and all times, although occasionally obscured by human weakness." Human reason was always subservient to divine revelation.[10] Hengstenberg was appointed professor extraordinarius in theology in 1828 at the University of Berlin.

Rose stated, very clearly, at the start of his series of sermons that his aim was to prevent "the Rationalizing school in Germany [ever operating among us]"; the German method subverted the divine by emphasizing "human understanding alone." They lived, he claimed, in an unprecedented time with "its tendency to exalt and exaggerate the powers and capacities of the human mind." In consequence, the veracity of the Christian system was now gauged by human reason alone and treated like any other scientific object. His own alternative view was that "the theologian must *mine* for the long hidden, treasure of truth, and like the naturalist must make new discoveries, and modify his belief accordingly."[11]

Rose's emphasis on the apostles, Articles, Common Prayer Book, and Anglican ritual naturally delighted Christopher Wordsworth and most of the High Church, but for his fellow members of the TC, at least now, Rose had totally misunderstood biblical criticism. In their eyes, German biblical scholars and philosophers were far freer to scrutinize the veracity of authorship and draw on historical sources and testimonies than the English.[12] However, Whewell would soon come much closer to Rose's stance, and Thirlwall by the 1860s would, ironically, bemoan the impact German biblical criticism had made on the Established Church. Rose's early fears, it seems, had proven right, and Thirlwall now acknowledged that such an approach should be severed. However, that was yet to come.

"KNOWLEDGE IS STILL, AS OF OLD, A HOLY THING": THE TRUE CHRISTIAN CHURCH

True Christian religion needed, Rose gravely warned, a "check and restraint" against destructive speculation such as German higher criticism.

The institution to do this was the Church of England and its ministers—"the fountain from which the living waters of truth are to flow clear and unsullied to the community at large." Consequently, the Church had to ensure it possessed "a clear and distinct declaration of faith, to which strict adherence must be required, of a liturgy which shall practically apply the doctrines of that declaration to men's wants and infirmities." The 39 Articles, enshrined in the Common Prayer Book, were to be obeyed. And it was the task of the Church, supported by the state, to "diligently repress every tendency to carelessness, and every attempt at innovation." German Protestantism had nothing sacred or permanent left and was at the mercy of fashionable philosophical whim: "Such views could not be held for a moment by those who had any belief in the divine origin of our religion." The Church of England, by contrast, "receives only what was received in those ages when truth must have been known"—and the stronger this faith, the greater morality and attendance of duty by man. How can there be any faith when "we fix in our minds to-day what to-morrow may teach us to reject?" How can the sinner be saved if the doctrines of redemption, revelation, the Holy Spirit, and the Resurrection are all "mere speculative fancies"? The truth of Scripture came from the divinely inspired voices of the apostles and their testimony should not be, and could not be, questioned by mere human reason. Rose, like the later Oxford Movement, insisted on the truth of the apostles and the apostolic succession; the Bible was the authoritative word of God because the apostles were inspired by, indeed, they *were* the voice of God. The German historical system of hermeneutic analysis was therefore "most fallacious and dangerous" and simply asserted, like French algebraic analysis and human logic along with, later, evolutionary natural selection, superficial human reason over godly authority. The lessons of mischievous German hermeneutics, deluded French mathematicians, dangerous deductive economists, and later, wrong-headed natural historians had no place in the realm of morality and higher truths.[13] By contrast, Thirlwall and Hare at this moment wanted to remove any human-based testimony from the true divine voice of the apostles.

So at one extreme all the Scriptures were inspired by God and then, at the other—far beyond the stance of Hare and Thirlwall—the opposite view that they were a human document. If the latter was true, the bolstering of the prevailing establishment via the teachings of the Church was also just a human construction and therefore could be likewise legitimately crushed. This was associated with English reformers such as Mill and Bentham. In the middle, contrary to Rose's allegations, was Thirlwall's and Hare's use of the German School to confirm the true divinely informed testimonies from the false human constructs. Whereas Rose saw any questioning of the apostles as blasphemy, Hare and Thirlwall saw it—at least during this moment—as making Christianity more robust and trustworthy. They all, however, had the same objective of preserving the Church of England and the social establishment. Hare's German approach to history probably went against him when the university appointed the staunch follower of Simeon, Christ's Hospital–educated and Trinity fellow James Scholefield as the new Regius Professor of Greek, after the former professor and Trinity don Peter Paul Dobree died suddenly and unexpectedly in 1825.[14]

For Whewell at this point Rose's stance toward human reason had gone too far. He had told Rose this just over a year prior to his rant against German higher criticism and underlined that some aspect of human reason was necessary. Whewell agreed that human reason alone stood far below Holy Scripture and true knowledge should, ultimately, be guided only via revelation. However, not all human thinking undermined Holy Scripture: "You are like St. Peter who would be crucified with his head downwards . . . why will you not see that in speculative matters, though reason may go wrong if not guided by our better affections, that you cannot do without her."[15] All our faculties needed to work together. Moreover, at this point Whewell was still optimistic about "scientific knowledge" and wrote to Rose, "I am persuaded that there is not in the nature of science anything unfavourable to religious feelings, and if I were not so persuaded I should be much puzzled to account for our being invested, as we so amply are, with the faculties that lead us to the discovery of scientific truth."[16] What Whewell meant by "faculties," though, was very different to many of his onetime Cambridge colleagues such as Babbage and Herschel.

Signs that Whewell was slowly moving toward Rose's stance could be gleaned later that year: "As to your attack on experimental science [including scientific German biblical scholarship], I dare say that when you had explained all the angry language that you thought it fair to use, when you found her poaching beyond her own domains, we should not be very wide apart." He also tried to be clearer about his view of science: "What I do hold is that inductive science is a good thing, and, as all truth is consistent with itself, I hold that if inductive science be true it must harmonize with all the great truths of religion." Moreover, "it is by no means correct to consider advances in science as rejection of what was known for something new. The novelty, if the philosophy has been duly inductive, including old truths and shews them from a new point of view."[17] Reconciling the history and philosophy of science to Christian religion, defined by the Church of England, was Whewell's role on earth.

Understanding the natural world and humanity, from Whewell's mutated form of Germanic idealism, all stemmed from the same divine creative will. If there was truth, it had not just suddenly appeared to this generation, but rather a great deal of truth already existed and was embedded in us, society, institutions, and the Constitution. By middle age Whewell's organic conservativism was complete. Providence, if God had given us the means to discover truth, had also allowed former generations: "The principles on which man's well-being, as to his soul, depends, have not been hidden in the dark till now. . . . We may have our new portions; we may add our share of struggle and progress; but we are but one generation, many have gone before."[18]

A sign of Whewell's gradual convergence toward Rose's position could already be seen in his response to an attempted defense of German theology by Karl Gottlieb Bretschne: "I did not think it very formidable for you."[19] A further indication came in a sermon also delivered in 1827 in which he denounced those who claimed that a few simple scientific axioms and a degree of "analysing sagacity" could, without the aid of Scripture, discover "what determines the happiness, the good government, the freedom of states; what secures the present administration of law; [and] the most desirable distribution of property."[20] This eradication of religious feeling had led to "mutilated and imperfect" knowledge. For example, those natural historians of the earth who have found "marks of revolutions and successions" have completely failed to understand that they were a product of a purposeful "Divine agent." The result was the corruption and perversion of the mind: "*Knowledge is still, as of old, a holy thing*."[21] For example, Whewell argued, "The great deluge was an event which however gigantic was yet small compared with those which proceeded the existence of those animals to which Adam and the sons of Adam have given names. And yet this event was made to produce consequences in the vital powers of man which we cannot conceive or understand."[22] Again, Whewell's emphasis on the Fall of Adam and subsequent corruption of knowledge was a very old and traditional argument.[23]

The TC rejected any linear notion of progress or, as was then coined, "the march of mind." The Hare brothers concluded their *Guesses at Truth* with the claim that it was only via historical study, not contrived Scottish stage theory, that "the gradual unfolding of all the faculties of man's intellectual and moral being . . . under every shade of circumstance, and in every variety of combination" would be revealed. The whole "*purpose and end of the history of the world is to realize the idea of humanity*." Humankind was created by God, was alive and growing. The object of study should also be nations, international law, and their differences over time.[24]

This was exactly Jones's approach to political or, as he, and indeed Whewell, preferred, "ethical economy." Attacking human-induced materialism and utilitarianism was a shared objective of an otherwise internally divided Church of England. For Thirlwall and Hare, this needed a more encompassing Church, while for Rose and many other High Anglican Churchmen of his ilk, it needed enforcing by a stricter, narrower, and more tightly policed National Church. By this point Whewell was with Rose.[25]

HISTORY, LANGUAGE, AND CLASSICAL ANTIQUITY

Much of the TC's fears and objectives was most clearly revealed when it came to understanding the classical era. The history of the Greek and Roman world was revised by new approaches in the works of German scholars such as Niebuhr, August Boeckh, Karl Otfried Müller and, of course, Bunsen.

The German historical method, especially through Niebuhr and his *Romanische Geschichte* (1811–1832) and *History of Rome* (1828–1842), has in certain quarters been acclaimed as an intellectual watershed for initiating a new style in the writing of history. The Danish-German statesman, banker, and historian dwelt on the critical evaluation of sources, especially of traditional accounts and the attempt to demonstrate the utility of a philological approach using language itself as evidence. Niebuhr further maintained an "evolutionary" interpretation of the history and development of all human societies, under which practical comparisons might be drawn between nations and epochs. Through his "Critical Method" he set out to analyze written accounts that had survived from the past and subject them to a philosophical investigation, which would unlock an inner meaning that had faded over time and was subsequently no longer suspected. He wanted to show that you could, contrary to most philologists, trace Roman history back beyond the third century BC via oral tradition preserved by poets.[26]

G. F. McFarland has described Niebuhr's approach as inductive but heavily reliant on intuition.[27] This latter component was not available to everyone and is a linking factor in Thirlwall, Hare, and Whewell's philosophy of understanding—particularly the notion of guessing and fundamental ideas. Whewell told Jones that an important aspect of his philosophy of induction "is the assertion of ideas and facts as equally and conjointly necessary to science—also the opposition between the indistinct knowledge which is implied in the use of common language, and the distinct knowledge for which induction inquires;—also the attempt to shew that guessing (and therefore often guessing wrong) is a necessary part of the progress of induction."[28] Guessing, using the imagination, was a vital part of induction, and this lay at the heart of his and Hare's philosophy. Throughout the early 1830s Whewell attempted to define his notion of fundamental ideas: "I expect to shew *clearly* that in order to arrive at knowledge or science we must have, besides impressions of sense, certain mental bonds of connexion, idea relations, combining modes of conception, sciential conditions, or whatever else you can help me call them: they are what I called Ideas in my former letter." "*Space*," he claimed, "is the ideal relation on which the science of geometry depends; *time*, *cause*, *likeness*, *substance*, *life*, are relations on which other sciences depend." Geometry was the tool to exhume space and time. Whewell hoped when he had discovered all the ideal relations of the physical sciences, he could do the same via analogy with geometry (or mechanics) with morals.[29] This, he hoped, would refute and replace political economy with a moral economy.

Whewell's mental struggle during the 1830s would be resolved and become dogmatically defined by the 1850s For Mill all this was simply an epistemological device to prevent political reform. As he convincingly wrote, "We do not say the intention, but certainly the tendency, of his [Whewell] efforts, is to shape the whole of philosophy, physical as well as oral, into a form adapted to serve as a support and justification to any opinions which happen to be established," that is, the Established Church, state, and prevailing sociopolitical order.[30] However, Mill's opponents had a counterargument, namely that the alternative could equally be equated to despotism, as demonstrated in "rational" France. False ideas based purely on abstract reason lead to false opinions, administrations, institutions, and morality.[31]

Niebuhr's work was a reaction to the "rationalist" and "conjecturalist" schools. The latter approach drew, partly, on a British empiricist tradition that posited that all knowledge was based on sense experience. As such all persons experienced the same physical world and thus assumed there was a psychological unity in humankind. For Niebuhr and subsequently the TC, this had no organic/historical foundation and, instead of nurturing moral/religious development, induced the opposite, namely moral decay, false knowledge, and an erosive disrespect for established authority. Very early on Thirlwall had condemned Locke as "unphilosophical" and the manufacturer of a doctrine that debased "our nature." Locke shook "the very foundation of our hope and faith in suggesting to us the fallacy of our most powerful convictions." Six years later Thirlwall wrote to Bunsen of his grave concern regarding the future of Albion: "The want of a right direction of intellectual energies is perhaps the greatest that England now suffers," with the country "exposed to all the perversities and extravagancies of powerful minds which have broken loose from the old arbitrary restraint, and are governed by no self-acknowledged well-grounded principles."[32]

No wonder Anglicans such as Whewell condemned views contained in, for example, the later anonymous (Robert Chambers) *Vestiges of the Natural History of Creation* (1844).[33] The argument of progressive evolution on earth without a continually intervening God was sacrilege. God, according to this claim, had set natural laws but then departed and certainly did not intervene via miracles. Whewell refused to review the book and instead gave his indirect response in his publication *Indications of a Creator* (1845). He told the sympathetic Reverend Frederic Myers just what was at stake: "I have attempted to show that, dim as the light is which science throws upon creation, it gives us reason to believe that the placing of man upon the earth (including his creation) was a supernatural event, an exception to the laws of nature. The *Vestiges* has, for one of its main doctrines, that this event was a natural event, the result of a law by which man grew out of monkey." For Whewell, this was not just ridiculous but profoundly immoral and distasteful.[34] Nonetheless nothing was ever quite the same and the mid-1840s was another nodal point in ontology and social relations.

Myers and Whewell were connected, not just through an education at Cambridge (Myers went to Claire), but via their wives. Myers was married to Susan Harriet Marshall and Whewell to Cordelia Marshall both daughters of the linen tycoon and one of the MPs for Yorkshire John Marshall. The families would often meet in Cumbria on Myers becoming the first curate of the newly built St. John's church in Keswick. Two of Myers's three sons, Frederic William Henry Myers and Arthur Thomas Myers, studied at Trinity. Frederic went on to be a Trinity fellow and a gifted poet, classicist, and philologist.

Here the Myers boys would have been taught by Whewell that the key element to truth was the relationship between "fundamental ideas" and the essence of real things. As Whewell wrote in *On the Philosophy of Discovery*, "The world is constituted according to the ideas of the Divine Mind, and the human mind apprehends the inward and most fundamental relations of the universe by sharing in some measure of those same ideas." These fundamental ideas embodied *an original divine language* containing an objective relation with what they described.[35] In one chapter of the *Vestiges* the author reiterated the unsettling argument that the original state of man was barbarous and not civilized and that language had evolved, in a Lockean sense, as a human tool and was not of miraculous origin. Whewell, in a short section on the "Origin of Languages" in *Indications of the Creator*, made it clear that the beginning of language was, on the contrary, not human, but divine in origin.[36] In general, Whewell moaned to Jones, "No really philosophical book could have had such success [as *Vestiges*] and the very unphilosophical character of the thing made it excessively hard for a philosophical man to answer it, and still more, to get a hearing he did."[37] Hence, of course, Whewell equated philosophy as religious and not human. However, over the following decades the opposite would come to be.

For the TC the foundation of mankind and language stemmed from an act of God. Thirlwall and Hare had over a decade earlier jointly published a translation of Niebuhr's *History of Rome* (1828) to demonstrate this. History was infused with God's providence, and its study thus provided a clearer understanding of the Creator. Niebuhr's history became an important text at Cambridge and Oxford, with a similar reading informing the work of Hare's brother, Augustus, at the latter.[38] Public reaction was mixed. Hare was furious after negative comments attacking Niebuhr appeared in the *Quarterly Review* by the second secretary to the Royal Navy and member of the highly utilitarian and empirical Astronomical Society of London, Sir John Barrow, a superintending clerk by training at a Liverpool-based iron foundry. Consequently Hare quickly set his pen to work to vindicate Niebuhr's credibility. Barrow had linked Niebuhr to the rioting of German university students and questioned the actions of Thirlwall and Hare for making the translation of such a potential troublemaker.[39] Hare condemned as nonsense the idea that the work nurtured a spirit of insubordination in students with, he reminded his readers, the first edition of Niebuhr's work being dedicated to the King of Prussia.[40] Hare wrote, "I know of no philological work, unless it be [Friedrich August] Wolf's [1759–1824] *Prolegomena ad Homerum* [1795], which has effected so much within the same space of time: it has flung an entirely new light over the whole region of ancient history; and hardly a treatise of any value on questions connected therewith has appeared during the last fifteen years in Germany, but we may trace the influence of Niebuhr upon it."[41] Hare and Thirlwall's translation became a firm favorite at the elite Anglican school, Rugby, under the auspices of its new headmaster, Thomas Arnold.[42] The feeling was mutual, and Hare later concluded that Arnold was a man of "extraordinary purity and holy energy."[43] Thirlwall completed his own *History of Greece* in eight volumes between 1835 and 1844. This, along with Arnold's three-volume unfinished *History of Rome*, written between 1838 and 1842, was probably the clearest example of Niebuhr's impact on British historiography at this time. However, both their works would soon be surpassed, according to Thirlwall, by his old Charterhouse friend, George Grote, and his two-volume *History of Greece* (1846).[44]

At the heart of Hare's and Thirlwall's defense of Niebuhr was an organic critique of radical change built on unprecedented foundations: "Niebuhr's political creed, it is manifest, classes him, now as formerly, among those who maintain the historical side of the question in the great controversy as to the first principles of government, and in opposition to those who would refer all such matters to the absolute decrees of the reason." Crucially, Hare concluded,

> he is with [Edmund] Burke, and against the *Contract Social.* There is nothing he would deprecate more earnestly than the cutting off those numberless fibres, by which the vital spirit of the past circulates through the present: he would have us go on inhabiting, and repairing, and, as occasion arises, enlarging the castle our fathers have left us, with its lordly towers, the regal battlement that crowns it, and the sacred chapel nestling in its heart; and would forbid our pulling it down in the presumptuous conceit that we may build a better from the ground.[45]

One problem was the fact that next to no primary documents existed from the ancient world, and consequently Niebuhr had constructed a history on very little evidence.[46] This was fine if you were guessing at truth but not if you were a historical empiricist. This was most powerfully made in the later stinging criticism of Niebuhr made by the historian George Cornewall Lewis in his huge *Inquiry into the Credibility of the Early Roman History* (1855).[47] This Etonian and Christ Church graduate had, ironically, once contributed to Thirlwall, Hare, and Rose's *Philological Museum* (1831–1833). By this stage, for Hare, Lewis had clearly lost his imaginative powers and fallen from grace. Moreover, he had become a key member of what, for the TC, was the revolting Society for the Diffusion of Knowledge.

Such criticisms were simply wrong. Hare reminded his readers that at the peak of revolutionary French power and influence, it was Niebuhr who had taught Prussia "the crushing tyranny of Jacobinism."[48] This great German hated the French revolutionary movement "because it has no reverence for the majesty of nature, but would level the distinctions of character, and substitute one vast bloody blot for the rich and varied hues with which the world has been adorned by its maker." French radicalism, as well as British utilitarianism and other rational sciences, had "no reverence for the majesty of history, but would render asunder the golden threads by which the present and the future are connected with the past."[49] The harbingers of the French Revolution believed true history had only now just begun. This wave of radicalism was socially dangerous and destructive: "Instead of revering ancestral institutions, we idolize modern abstractions, and lose our individuality in a cosmopolitical indifference."[50] It would level society and social distinctions through the tyranny of human democracy.

In contrast to France, Niebuhr had much earlier claimed in an article published in the *Prussian Correspondent* that the true light had been revealed in England:

> The peculiar manner in which the Reformation was affected in England, has led the Episcopal Church to preserve a very great number of excellent institutions, which have unfortunately been destroyed in the Lutheran and Calvinistic churches. He who deplores that these churches have lost the sublime ritual and ceremonies of primitive times, and whose feelings and imagination are attracted by the worship of the Roman Catholics, would find, if he were acquainted itself with the English episcopal church, that it satisfied all the wants of his heart; nor has it separated itself from the fathers and the whole body of the primitive church, like the other protestant congregations.

For Niebuhr this preservation was to be the virtuous source of the country's defeat of the French during the Napoleonic Wars and the rise of Britain's unprecedented hegemony. History commenced with the divine Creation and not the human-created French Revolution. Founding a world on purely human abstract principles was godless and a stark contrast to the ongoing historical and organic institutions of Britain.[51] Coleridge had earlier emphasized the older the language, the purer it was; words were alive and mutating from an original divine source—a collaborative process between imagination, guessing, and human reason. The sense and meaning of a word could be grasped only by constructing some form of historical dictionary.[52] Thence the importance of etymology and philology to the TC and numerous others.

Hare's good friend, Arnold, placed the study of classical languages, with a German bent, as central to his curriculum at Rugby School. A controversial move since, as we have seen, many saw Niebuhr as blasphemous and dangerous. Thirlwall was at a loss and told Bunsen,

> [The] irreligious atmosphere [in England] is a very peculiar one, as may be supposed in unknown tongues, which are now attracting crowds to one of our meeting-houses. The millenarian persuasion is become so universal that any man who doubts the certainty of the Messiah's appearance on earth being now near at hand is denounced by, I am afraid I may say, a majority of the persons who claim the epithet religious by way of eminence as a downright infidel. That persons of this description would be scandalized by Niebuhr's divergency from the book of Genesis I knew to be an unavoidable misfortune, and I only hoped his speculations might not fall into their hands.

The Church of England was too occupied in preserving its increasingly perilous relationship to the state, confessed Thirlwall, to embrace the application of German higher criticism to theology. He fully supported the need for an established form of worship but felt this desperately needed a "revision of her liturgy"—a reformation that would be broad enough to encompass religious Dissenters.[53] By the 1850s, however, Thirlwall's stance would be radically altered.

THE CAMBRIDGE ETYMOLOGICAL SOCIETY AND PHILOLOGY

In 1831 the TC formed the Cambridge Etymological Society (1831–1832) and its accompanying journal, the *Philological Museum* in a deliberate attempt to broaden classical studies. This short-lived publication was set up to further the methods of German scholarship and was modeled on *Rheinische Museum* (1820s), edited by Niebuhr, and *Museum Criticum* (1814–1826), edited by the former Trinity fellow, teacher of Thirlwall, and Regius Professor of Greek, Monk.[54] Prominent contributors to the *Criticum* had included Blomfield, yet another fellow of Trinity.[55]

Monk and Blomfield had been students and followers of the Etonian and Trinity-educated, Richard Porson, Monk's predecessor as Regius Professor of Greek. They tended to publish works on Greek drama and emphasize grammar, syntax, and meter. In addition to publishing articles by Porson, they also introduced notes on the earlier Trinity fellow and master of the college Richard Bentley. As we have seen, Monk subsequently, in alliance with John Kaye, master of Christ's, and, of course, Christopher Wordsworth, became instrumental in establishing the Classical Tripos in 1822.[56] In contrast to the *Criticum*, the *Philological* was much more an outlet for the new comparative philology being developed, especially in Germany. The journal was spearheaded by Hare and Thirlwall. Christopher Stray sums up their quest: "They wanted the empirical knowledge of the classical languages characteristic of the English school to be replaced by scientific investigation of the classical literatures as pursued in German universities."[57] An objective Thirlwall would come to deeply regret.

Niebuhr was a key figure in Hare and Thirlwall's intellectual approach. He defined philology as "the *introduction* to all other studies [preserving] an unbroken identity through thousands of years with the noblest and greatest nations of the ancient world, with the work of their minds, and the cause of their destinies, as if there were no gulph dividing us from them."[58] Illuminating this historical consciousness through the historical study of language would, he believed, reveal the single divine spirit.[59] Each historical period was to be treated within its own context and the prevailing expression of peoples; knowledge of the divine was different within its epoch but not necessarily wrong or inferior—a natural reflection of the context's consciousness at the time.[60] This was also, of course, central to Jones's and Whewell's ethical economy.

Hare and Thirlwall both aimed to scrutinize the whole of ancient literature as a way of rediscovering the fundamental essence of ancient Greek and Roman civilization, indeed, universal aspects of humanity. This demanded a comparison of languages, and Hare at least seems to have inherited the skills of his uncle, Sir William Jones, who had spearheaded comparative philology, in particular the relationship between Sanskrit, Greek, and Latin.[61] Thirlwall also pioneered in Britain the importance of material evidence to history, especially the use of inscriptive evidence as a tool to illuminate past eras and gauge their level of civilization. Even more important was the need for a comparative approach.[62] This last point was also a key tool in Richard Jones's attempt at demolishing Ricardian political economy.

The *Philological Museum* never sold well, and the magazine died when Hare left for his church living in Sussex and Thirlwall resigned from Trinity in contumely. It was therefore left to a version of William Jones's comparative philology to take over from the philology of ancient European history, in particular his argument concerning the relationship between Sanskrit and classical languages as developed by German scholars such as Franz Bopp, Schlegel, Wilhelm Carl Grimm, Alexander von Humboldt, Bunsen, and later in Britain Max Muller. The emphasis now was on shared grammar rather than etymology. As J. W. Burrow writes, "The revelation that Sanskrit was the elder sister of the classical and Romance languages, and, as it later appeared, the Teutonic as well," meant that "the classification and derivation of languages could be put on a systematic footing and linguistic change could be studied on a comparative basis." Language was not just some crude human sensory reaction to a person's surroundings, but something deeper within human nature; it stemmed from God but had subsequently mutated and needed purifying.[63] We had to free ourselves from mere sensory appearances and mediate inward on our consciousness. Following Kant, and of course Plato, the focus should be on ideas and not images.[64]

The TC's linguistic quest was shared by several others. James Cowles Pritchard, for example, was keen to use philology to map the prehistory of ethnic races. Following the spectacular surprise that the languages of India and Europe were linked, the project seemed eminently possible. A related debate was raging too over the origin and geography of objective measures in metrology: Was it ancient and divine in origin?[65] Despite coming from a Quaker background, Pritchard trained to be a doctor in Edinburgh and gradually became more and more Anglican in orientation. He subsequently spent a few terms at Trinity from 1809 before going to Oxford, where he converted to the Church of England. His biographer writes, "Pritchard's main ethnological aim was to prove the unity of mankind and to vindicate the veracity of the anthropology laid down in Genesis . . . he ventured into philology to show that all human languages were related, whence he inferred genealogical relationships among human tribes."[66] In 1813 he published his two-volume *Researches into the Physical History of Man* and a second edition in 1826. Pritchard claimed that all earth's languages had originated from "one stock," but the traces to this common origin were now missing. This argument was appreciated by Whewell and by Anglican evangelicals advocating the liberation of slaves. Everyone had originally descended from Adam, had inherited original sin, and consequently needed saving—be it via missionaries in Africa or university dons at Trinity.[67]

Another well-known philologist was the King William's College–, King's College London–, and Trinity-educated Frederic William Farrar. During his time in London he came under the spell of a close follower of the TC, Maurice, then professor of English at King's College. Farrar also became a member of the Cambridge Apostles, won the chancellor's medal for poetry in 1852, came in fourth in the Classical Tripos in 1854, won the Le Bas English essay prize and Norrisian divinity prize for an essay on atonement, both in 1856, and finally was made a fellow of Trinity that same year. Maurice had motivated him to read the complete works of Coleridge—Whewell was deeply impressed.[68]

These philologists all admired the work of the Trinity fellow John William Donaldson. In 1830 Donaldson had gained the first prize in Greek at University College, London, and subsequently migrated to Trinity. There he became part of the Hare and Thirlwall circle and managed to come second in the Classical Tripos and senior optime in 1834. In 1839 he published his comparative philological work, *The New Cratylus, or, Contributions towards a more Accurate Knowledge of the Greek Language*. The aim was to utilize philology to expand the study of Greek and Latin. Donaldson resigned as a fellow in 1840 after getting married and becoming the new headmaster of Bury St. Edmunds grammar school (King Edward VI).[69] This, it will be recalled, was the same institution Blomfield, Spedding, Kemble, FitzGerald, and Donne had attended. Donaldson succeeded their headmaster, Malkin, after the latter had been appointed the first professor of history at University College, London.

Donaldson argued that language was perfect at the beginning of time and had subsequently been corrupted by the "increased use of inflected language" (the modification of a word that, in turn, changes the grammar). *The New Cratylus* specifically aimed to refute Tooke and elevate religion and morality—precisely the objective of the TC. Thus the battle was once again against the rise of philosophical materialism and the idea that humans had evolved from a simple state of savagery. History of the classics—especially the Bible—would demonstrate that revelation was an ongoing process.[70]

As already noted, the Cambridge Etymological Society made something of a comeback in 1843 with the formation of the Philological Society of London. This had an initial membership of 203, with Thirlwall as president until 1868, when he was replaced by Thomas Hewitt Key (Trinity). Hans Aarsleff writes, "A detailed examination of the list of original members, their education and earlier history, will show a very close connection with Cambridge, and especially with Trinity College." Three out of Thirlwall's vice presidents were Cambridge men: Robert Gordon Latham (King's), Bishop Blomfield of London (Trinity), and Bishop Maltby of Durham (Pembroke). Other Trinity members included Whewell, Hare, W. H. Mill, and St. Paul's–educated James Prince Lee, now headmaster of King Edward VI's School in Birmingham. Elsewhere from Cambridge were the Etonian Edward Craven Hawtrey (King's), now headmaster of Eton, and Donne.[71] Yet, despite the strength of this new London group, Whewell suggested to Hare that when they grew old and were released from formal duties, "we might re-establish the Etymological Society and work out some of our old plans? I do not see the philological Society is likely to supersede our labours."[72]

As we have seen, a prominent target of the TC and other philologists was the nominalism of English etymologists like Tooke and, more generally, utilitarian exponents of the new political economy, such as Bentham, Mill, and Ricardo. When it came to Tooke, Latham sarcastically smirked, "No system of materialism will account for grammar, that form of language which is due to the pure reason."[73] Elsewhere Whewell and Jones brought the historical method to counteract what they believed was the destructive abstract deduction dominating a growing self-confident political economy seeking the status of a science; this movement propagated false laws. Deductive reasoning devoid of God was simply erroneous. Whewell published two papers on political economy in 1830 and 1833 purporting to mathematize the subject, but with the ironic aim of showing that such an objective was impossible.[74] Whewell's "object" was "to shew the mode of applying mathematics so as to separate difficulties of calculation from difficulties of moral reasoning—to keep apart the business of reasoning up to principles & down from them."[75] He hoped that his work would "prevent my friends the young mathematicians from looking with too profound and unquestioning an awe upon the crabbed dogmatizing of the millions."[76]

Whewell knew he could not prevent the tyranny of utilitarianism and democracy, but he hoped to be able to keep them in check. He told James Marshall, "If we can find social, moral, and religious truth, we must disregard the voice of the strongest and largest multitude when it utters falsehoods."[77] In other words, the tyranny of democracy stemmed from corrupted human reason. During this period Whewell also found time to police scientific terms. For example, he was responsible for Michael Faraday's adoption of electrochemical terminology derived from Latin terms like *electrode*, *anode*, *cathode*, *electrolysis*, and *ion*.[78] As Robert O. Preyer comments, "If knowledge is power, then the prolific coiners of new words are the power brokers of intellect." Whewell took it on himself to be the authority of new terms in natural philosophy such as the above and, for example, in geology. In the latter, for instance, he coined the terms *uniformitarianism* and *catastrophism*; moreover, he established the words *scientist* and *physicist*. Language played an important part in discovery and was the link between fundamental ideas and sensation. Whewell was keen to develop a scientific nomenclature that would retain a "root" concept and hook additions on either side to allow an ongoing notation; a glue of otherwise random facts to a general idea. This would ensure it retained a recognizable form. Preyer points out how this "definition-by-usage" approach was adopted by the editors of the new *Oxford English Dictionary* in 1884. This emphasis on etymology was also very characteristic of the work of Coleridge.[79]

It was Jones's task to publish and demonstrate what they deemed was the correct method for understanding the economy. The debate did not simply involve intellectual issues concerning the status of method but in general the structure of politics, nature of morality, and appropriate institutions. Reform based on false principles had catastrophic implications. At stake was the nature of the past, present, and future, which had serious repercussions for institutions such as Trinity, the British Constitution, the Church of England, and traditional social authority. Whewell utilized language as a tool to police and manage the establishment of "permanent knowledge, and appoint those who were authorised to introduce, change and preserve this terminology." This fundamentally included safe Anglican policed science.[80] After all, he truly felt he had cultivated his soul, had the potential ability to access a glimmer of God's original language, and correctly documented the divine-induced path of scientific discovery. No wonder philology was so important to him, along with Hare and Thirlwall.

One former, anonymous pupil of Hare's at Trinity reminisced about his master's philological teaching and the importance of understanding the essential importance of tenses, nouns, verbs, and grammar. These had to be obtained to gain access to the "divine intuitions of the poet." He did not know exactly how Hare "imparted this conviction" to him, but "one rose to the apprehension that the poem *had* a unity in it, and that the poet *was* pursuing an ideal, and that the unity was not created by him, but perceived by him, and that the ideal was not a phantom, but something which must have had a most real effect upon himself, his age, and his country." Hare's intention was not, the former student reflected, to teach us morals but to show us how we might find them within us: "Thanks be to God who has forced me to acknowledge that there is an ideal, in which and after which man is created; an ideal which explains and justifies all the ideals men have perceived, and followed, and found themselves unable to reach; an ideal which tells us what our sin is; an ideal which can lift us out of it."[81] It was the quest to impart this ideal into first a Trinity, and then a Cambridge, education that unified the TC.

However, the privileged cloistered study of language was not diluting the surge for change. During this period of immense social and religious reform, the idea of a major transformation, if not disestablishment, of the Church was an acute fear. The impetus was coming from a growing number of Dissenters, Whigs, and atheists. Things became very real and dangerous when an incident occurred next door to Trinity. The ex-cons and blasphemers Richard Carlyle, a tinplate man by trade, and Reverend Robert Taylor, later dubbed "the Devil's Chaplain" by Charles Darwin, arrived in Cambridge on Thursday, May 21, 1829, to promote infidelity among the student populace. The next day they set up their headquarters at lodgings above the print shop at 7 Rose Crescent in the heart of Cambridge and just by Trinity Street.

By noon they had issued a printed challenge to the vice-chancellor, college heads, doctors of divinity, and Taylor's old Cambridge mentor, Simeon, advertising themselves as ready to debate the merits of the Christian religion and the nonexistence of Jesus Christ. Taylor, dressed in university apparel, walked through the town and colleges handing out leaflets and looking for converts or sympathizers to their cause. The backlash came on Saturday as the proctors swung into action. An article written for the local newspaper by Taylor and Carlyle failed to appear, while the proprietor of the lodging house they were staying in, Mr. Smith, had his licence revoked. An outraged group of students planned to treat the duo to a punishment for their apparent deception. Taylor and Carlyle apologized to the landlord, swiftly departed, and wrote to the vice chancellor to reinstate Mr. Smith's license.[82] The future of the Church and a Cambridge education took another turn for the worse, especially for Thirlwall, during the early 1830s.

CHAPTER FOUR

TRINITY AND RELIGIOUS DISSENTERS

To be sure, the TC was divided over certain prominent issues. As we have seen, Rose began to look in revulsion at the implications Germanic study could have on the Established Church by questioning the gospels. He was one of several prominent High Church Anglicans who emphasized the veracity of Scripture and the need to defend it from those who sought to undermine its divine origin. The Bible was the word of God—therefore beyond scientific scrutiny and not limited to a historical context. Where Hare and Thirlwall at this point sought a wide National Church that could accommodate different views (Anglicans as well as Protestant Dissenters), Rose and his ilk sought a more narrow and authoritative Church that kept religious Dissenters out. Any critical inquiries of Scripture and any questioning of divine inspiration had to be rejected. This tension came to a head at Trinity in the early 1830s.

LEGISLATION, RIOT, AND POLITICAL REFORM

The rude arrival of Carlyle and Taylor occurred just as proposals were being made for political and Established Church reform. The fierce utilitarian John Wade had sold fourteen thousand copies of his *Black Book, or Corruption Unmasked! Being an Account of Persons, Places, and Sinecures* during the 1820s. This was a hostile attack on the establishment and National Church. He republished it in 1831 as *The Extraordinary Black Book* and sold a further fifty thousand copies by 1835. The book caused a public uproar. The Test and Corporation Act of 1673 had been repealed in 1828, which now meant anyone holding public office in Britain need not subscribe to the articles of the Church of England. The following year it became legal for Catholics to take a seat in Parliament with the passing of the Roman Catholic Relief Act. The threat of civil war in Ireland was also a real fear. When cholera hit Britain in 1831 it was seen by many as a judgment from God against these constitutional reforms;

thirty-two thousand people died in 1831–1832 and frustration led to attacks on doctors. The following year saw the expansion of the franchise with the Great Reform Act.[1]

It has been estimated that as many as 250,000 religious Dissenters (including Catholics) gained the vote in 1832. Could the religious inequality of the Established Church continue to be sustained? Jones said he was "dismayed at the reconstructive part of the reform bill & think it weak & mischievous very contemptible & very alarming."[2] Whewell was equally hostile and wrote back, "I am resolved to possess my soul in place and to leave the people to pull down the social edifice for their amusement which at present they seem disposed to do."[3] Thirlwall was fearful for the Church's future. He accepted that some reform of the Church's liturgy was needed, but "decidedly prefer[red] an established form to an unlimited licence in the management of public worship." He further told Bunsen in 1832, "The service of the Church ought, according to my view, to be regarded as the highest work of art. If it does not purify and elevate the soul." With both eyes directed at Cambridge and Oxford he warned that if it failed in this capacity it was due "to the incurable apathy of the persons who take part in it or to the indolence and unskilfulness of those who conduct it."[4]

The Reform Act unleashed an array of legislation seeking to reform ancient institutions on utilitarian lines.[5] There was no respite, and the quest to restructure the Church intensified. In 1835 the Ecclesiastical Revenues Commission was formed to examine the Church's structure and administration.[6] Major topics for the modification of the Church included the commutation of tithes in 1836 and attempts to abolish church rates. Foreboding signs such as the refusal by Birmingham to pay church rates enhanced the pressure.

The Church rate was a tax on the parish to keep their church in good repair, with a victory for Dissenters in 1833 when the levy was abolished in Ireland. The state compelling Dissenters to pay for church repairs to a religion they rejected was hugely resented. Part of the income from the tithe was used to repair the chancel (the area of the church near the altar), typically reserved for the clergy and choir. Meanwhile to remain in power required the Whigs to actively continue to woo the support of religious Dissenters. This was seen by Tories as a cynical policy to pave the way to the eventual disestablishment of the Church. The tithe, for example, was revenue that ended up in the purse of the Anglican clergy—although many of the tithes from the monastic houses since the Reformation actually went to the landed gentry. This was an issue that got caught in the anger sweeping the country during the Swing Riots.[7] A few years earlier Whewell had received a Captain Swing letter that stated, "If you do not call in all the copies of your Mechanics which have promoted the building of Machines you shall hear further from Swing."[8] Whether this was a student prank or not, the irony would have made Whewell smirk.

Jones looked on in fear as impoverished agricultural workers in his home county of Kent vented their anger by destroying agricultural machines. Even more worrying for Jones and his kind was their increasing hostility toward the Church and tithes. Early in 1832 three large uprisings took place just over seven miles from his home—with another as close as four miles away. He had hoped troops would have crushed them, "& an example made which would have protected us & the poor people themselves for a generation at least."[9] Whewell understood Jones's concerns, for only two years earlier social protests had ravaged the vicinity of Cambridge: "We have had two fires—one large one so near as to illuminate our great Court." The Trinity "youths," Whewell triumphantly wrote, "have armed themselves with sticks and anticipate a row with much glee."[10] However, with all the changes at large, Whewell's reliance on a young Trinity army to prevent political assaults was looking increasingly precarious.

Jones, in his lectures throughout this period at the East India College in Hertfordshire, boiled much of this immorality, violence, and disrespect down to the relatively new commercial, urban and industrial society that now characterized a still predominantly rural Britain. The old parish system was not adjusting to the new world. Jones warned his students, "The crowded state of the towns, the turbulent character of the workmen, some gross habits and some serious vices which are connected with these, and occasional spectacles of appalling want in contrast with surrounding luxury,—these things discourage many pure and ardent minds." The problem transcended social barriers and was a product of the prevailing commercial spirit and selfish individualism. He continued, "The eager and bustling pursuit of wealth by the middle classes, the occasional insolence of vulgarity and riches, and the disgust inspired by seeing these in combination, sullying, as it were, the legislative bodies, and degrading the highest functions of public life, these things add to the unfavourable impression" characterizing modern Britain.[11] It was the expanding selfish, materialistic, and commercial class that was to blame.

Likewise, Thirlwall confessed to Bunsen, "Is the reform of the Church to be consigned entirely to politicians and economists, who only look at the goodly stones and gifts of the temple, some with an anxious, others with a greedy eye, and care nothing about the service of the sanctuary nor the edification of the worshippers?" Depressed by the failure of the

Philological Museum project, Thirlwall told Bunsen he thought England was following France along the path of reason worship, and that before long the study of ancient history and philology would be considered, as it was in France, "a kind of fancy not much more respectable than that of a bibliomaniac, and as an indication of a somewhat weak head."[12]

RELIGIOUS DISSENTERS AND TRINITY COLLEGE

Another grave issue was whether Oxford and Cambridge should admit Protestant Dissenters to their degrees. Trinity supporters for the change included Thirlwall, Airy, and Sedgwick. A broader Church would, they believed, strengthen it. Strong support also came from another Trinity don, MP for Cambridge and Cambridge professor of political economy Pryme. Whewell despised his work on the subject.[13] A Cambridge petition composed of sixty-two resident members of the Senate was sent to the House of Commons in March 1834. It was presented to the House by a Trinity man, the Whig politician Thomas Spring Rice. The following month another Trinity graduate, the Duke of Gloucester, presented the House of Lords with a counter petition signed by 258 Cambridge University members.[14] The latter told the Lords, "The principle on which the University of Cambridge was founded, and by which it was guided, was that it was for the education of persons of the Church establishment, and especially for the education of those who were destined to be the clergy of that establishment." He also added that it would be crazy to change the religious requirements of the university. The institution had prospered under it and was responsible because "Newton, who possessed the greatest mind that ever man had, and other men of great eminence, had been reared within the walls of the University."[15]

Smoke was added to the fire when Thirlwall published an open letter on May 25, "On the Admission of Dissenters to Academical Degrees," in which he condemned compulsory chapel attendance as, in carefully chosen words, a "heartless mechanical device." This was officially a reply to a defense of the status quo pamphlet on the subject by Thomas Turton of St. Catherine's, the former Lucasian Professor of Mathematics and now the university's Regius Professor of Divinity.[16] Turton was worried about the impact that differing religious views would have both on the morality of students and on the supply of clergy for the Church. He was supported by the master of Trinity and the latter's youngest son, a fellow and tutor at Trinity, also named Christopher Wordsworth. The younger Winchester-educated Wordsworth denounced the proposed changes in a letter to Lord Althorp on April 12. In response Thirlwall condemned Turton's argument as "utterly chimerical" and challenged the assumption that the university was an ecclesiastical establishment despite the fact it had not been an Anglican seminary for centuries: "We have no theological colleges, no theological tutors, no theological students," unlike in Germany where there were approximately seventeen faculties of theology. Nor was divinity seriously taught at Oxford and Cambridge as it was in Germany. Besides compulsory attendance at chapel was putting students off the Church by inflicting on them a mechanical and forced system of worship, rather than searching for God.[17]

Thirlwall assured Bunsen that the demand for admission of Dissenters was "very reasonable," while recognizing that "those who look not only to the thing in itself, but to its consequences, believe that such a measure would subvert the Universities as ecclesiastical establishments."[18] At bottom, of course, this was a dispute about the role of a liberal education. Reformers wanted the ancient universities to wrest back the cause of intellectual advancement and "scientific progress" from places like the new London University.[19] This aim was clearly incompatible with the views of those who believed universities were there just to inculcate divine truth and breed future clergymen and politicians.

High Anglican Churchmen were already gravely concerned about the German biblical criticism being developed by Thirlwall and Hare, and about the Oriel Noetics at Oxford. Thirlwall's support for Dissenters merely confirmed their hostility. Rose was especially furious, and significantly he had the ear of the master of Trinity. In fact, he had been out to punish Thirlwall ever since the latter's translation of Schleiermacher, and here was a perfect opportunity. Admitting Dissenters to degrees would unravel the establishment and erode the apostolic tradition. The one thing needful was to preserve the four gospels of the New Testament unchanged. Faith in Christ came from God; Christ gave it to the apostles, who then passed it on to the clergy.

In a carefully timed sermon on May 18, 1834, Rose preached in St. Mary's on "The Duty of Maintaining the Truth." He gravely warned the undergraduates to be suspicious of change and to respect the university's traditions and principles: "It is our happy lot to have, or to have had, the privilege of preparing ourselves for the duties and trials of life, in these quiet and holy places set apart by the wisdom and piety of wise and pious men, to be seminaries of religious education as well as of sound learning (God grant that they may ever continue so)." He cautioned against the current fad for social, political, and religious reform: "We destroy indeed—and that will be the characteristic of our age."[20]

Eight days later, with Rose's words still ringing in his ear, Wordsworth wrote a public letter addressed to the fellows of Trinity in which he encouraged Thirlwall to resign his appointment as assistant tutor in the Classical Tripos. How could he remain a college member when he differed so much with the principles of the college? "Your continuing to retain it would, I am convinced, be very injurious to the good government, the reputation, and the prosperity of the college in general, to the interests of Mr. Whewell in particular [Thirlwall was his tutorial assistant], and to the welfare of the young men, and of many others."[21] Even more menacing was Wordsworth's opening gambit in which he questioned how Thirlwall could even remain a member of the college, seeing how he differed so much with its principles. Thirlwall, in turn, addressed a circular letter to the fellows, asking each of them to send him "a private explicit and unreserved declaration" on this point. Most desired to retain him, but several—including Whewell with whom he quarrelled publicly on the matter—did not acquit him of rashness; a few did not condemn the master's action.[22]

The Trinity master subsequently found he was even more unpopular in the college and longed to resign. However, he would not do so until a period of Tory administration was installed in power. This was to ensure his successor was cut from the right cloth.[23] His former reference to Whewell was very telling, and he was clearly already thinking of him as his ideal replacement. Wordsworth was alienated from many of the college's liberal fellows due to his strict "creed of anti-Jacobin Toryism." There was nothing new about his obsession with the chapel. As soon as he had been appointed master he had met with Henry Coddington and Thorpe "on the subject of an increased attendance at Chapel, on the part of the Batchelors."[24] He subsequently increased the seating and had built new high chairs for the deans to sit upon and keep a careful eye on the students' behavior. The atmosphere must have been stifling.[25]

Hare was infuriated by the thought that Rose had instigated the crisis—if so, then the man "ought never to be admitted into any room in the college again"—but his prevailing feeling was one of sadness. "It is rueful to think of the breaking up of that singularly happy delightful society which we enjoyed for so many years at Trinity. But how could one expect that it would be privileged to last forever." Since the passing of the Great Reform Bill, "every institution" had been shaken and all former ties "throughout the country" loosened.[26] His anger was rather reserved for Wordsworth, and he expostulated to Whewell, "Surely this is a most outrageous step. The high church party seems all gone stark-mad, and to have been all seized with a fanatical desire of martyrdom at all costs and risks. Else I should be utterly unable to comprehend how the Master could be guilty of such a piece of insolence and folly."[27] As it happened the attempt to allow Dissenters to take degrees was eventually defeated in the Lords. Ironically, Hare's views had hardened by this time, and he had come much closer to Rose, Wordsworth and Whewell on the issue of chapel attendance.

Whewell's role in all of this was careful but clear and determined. He was a very close friend of Thirlwall and Hare but was also the master's favorite and intimate with Rose. He opposed the Dissenters' claims and believed that those who matriculated should be compelled to attend chapel, but that because of the personal delicacies he expressed himself diplomatically, at least to start with.

Whewell had earlier told Jones he was against Whig reformers who wanted to see unrestricted admission and graduation of Dissenters. The Whig petition for such a policy "appeared to me very wild, except as a mere ministerial move. As to the substance of the petition, it throws down before the Dissenters the college Fellowships, which they did not ask for, but which being thus offered to them they will of course claim. *I think the Fellowships a necessary support to the established church; and I think the church a necessary part of our social system*" (my italics).[28] He was thus disillusioned with the views of the senior Trinity fellow and future archbishop of York, Thomas Musgrave, and with Sedgwick, Peacock, and especially Thirlwall for their support of the petition. The fact that Hare, Musgrave, Sedgwick, Peacock, Joseph Romilly, Airy, and Sheepshanks all supported the petition must have left Whewell feeling very isolated.

Thirlwall explicated his response to Wordsworth's démarche in a revealing letter to Pryme:

> I may also add that I showed his letter to Whewell before I answered it, and that, though he [Whewell] expressed great regret, he did not intimate that there could be any doubt as to our connection being at an end. In my answer [to the master] I treated his language as conveying an absolute command, to which I professed to submit, not because I was satisfied as to his right, or because I deferred to his judgment, but simply because I could not resist his assumed authority without involving other persons, namely, Whewell and [Charles] Perry, in the struggle; I think this was the only course I could properly take. Even if I had stood alone it would scarcely have been possible to retain such an office in spite of the Master; but connected as I was with Whewell and Perry, I felt that I had no choice left.[29]

Perry was an ex-Harrow schoolboy who had graduated senior wrangler and first Smith's Prize man in 1828. He subsequently became a Trinity fellow before being appointed college tutor between 1832 and 1840.[30]

Whewell, however, could not let the matter rest. Despite his appeal to Sedgwick that he did not approve of the master's removal of Thirlwall from his assistant tutorship, the fact is that he clearly did think this. Rather, he feared a violent reaction from fellows and students and appealed to Sedgwick, who was popular among them, to dilute their feelings.[31] He became further immersed in the struggle by publishing an outwardly courteous but damning critique of Thirlwall's views. Daily services had a disciplinary effect on immature undergraduates. They fostered the "latent and silent agencies of habit, and association, and reverential feeling, as means of controlling and sobering the wayward and wandering impulses of human nature." He concluded, "Our course of study is not constructed for the year, it is not framed in an accidental and arbitrary manner, it is not that which one person, at a particular time, happens to have a fancy for; it is that of which the origin goes back to the origin of our Colleges." To tamper with it could bring the whole edifice down.[32]

The respectful language could not disguise the fact that Whewell was becoming more and more livid. Indeed, what he said in substance was much the same as Rose, who told the university to be on guard against reformers: "Learn here to despise, as ye ought, the solemn mockeries which tell you that Truth changes with times, and that principles must be altered to meet altered circumstances, that you must now bow to the voice of the million, and hold public opinion for truth."[33] Rose congratulated Whewell on his response to Thirlwall's pamphlets. They had been truly "mischievous in every way. . . . The theory is bad as you have shown—the feeling is bad, the tone bad & the arguments resorted not worthy of a gentleman."[34] Just before Rose died he wrote to Whewell, thanking him for his friendship, work, and "your usefulness."[35]

Rose at first worked closely with the Oxford Tractarians. Newman became a regular contributor to Rose's *British Magazine*, although there were squabbles over the alliance between Church and state and over ways to make the Church more popular.[36] Newman and Hurrell Froude, for example, felt this alliance corrupted the Church—whereas all the TC saw it as vital. Their Oxford tracts were designed to discover what bedrock Anglican doctrine historically was before it had become encrusted with all sorts of post-Reformation accretions. For these Oxford men, writes Owen Chadwick, "the Church of England claims the allegiance of the English people not because it is recognised by the state but because it is the divinely authorised teacher of truth. The dissenters are wrong not because they refuse a state establishment but because they refuse constituted *Christian* authority."[37] However, the Oxford men's increasing move toward the Roman Church led them away from Rose and the *British Magazine* to the *British Critic* as their mouthpiece. Nonetheless, Newman dedicated his fourth volume of sermons to Rose, and Rose, despite misgivings, continued to support him. For men like Newman, if antiquity was the basis of the church then there was no contest; it was the Roman Catholic Church—the roots of the Church of England stemmed from Rome.[38]

Whewell described Thirlwall as someone who had caught, rather like flu, the prevailing reformist zeal:

> [He] could not understand why the monuments of the same feeling in our habits and institutions should be supposed incapable of contributing to the same effect [religious thoughts]; why the pealing organ, the well-known liturgy, the long-established orderly distribution of times and offices, the collegiate congregation ever entering and issuing from the consecrated doors with the returning hour, should not also be favourable to the growth of religious thoughts; or why we should be utterly callous to those impressions, arising from such circumstances, by which strangers are often forcibly affected.[39]

To allow Dissenters in would be to subvert the whole character and history of Trinity and the Established Church. Compulsory Anglican chapel attendance was crucial to this continuum and the correct teaching of the Church's catechisms. The potential move was also part of a larger project by the liberal reformers to change the role of the universities and make them truly national institutions by severing the power of the individual colleges from that of the university. A broad church was also part of implementing a broad education.[40] This was something Whewell fought against most of his life.

Thirlwall, in a series of long exchanges, told Whewell that the "real difference of opinion between us turned simply on the value of our Chapel service, in a religious point of view: you rated it very high, I very low," but while the difference might have been simple it was also very deep. Thirlwall insisted that religious prescription against the will always "kills religion or religious feeling," whereas "voluntary exercises of religion tend to promote it." And when Whewell countered with the argument that religious truth had to be inculcated by college tutors at their daily services, Thirlwall airily replied that he did not consider himself "as having anything to do with the administration of chapel discipline, and was not conscious of having neglected my duty. . . . I conceived that my official duties were confined to giving literary instruction."[41] And when Whewell protested that there needed to be uniformity in the Church's doctrine and teaching because religious knowledge was "the highest knowledge," and "the State will require the teaching of other knowledge . . . to be consistent with religious knowledge,"[42] Thirlwall and Hare retorted that it was precisely its emphasis on strict uniformity that threatened to make the Church obsolete. Just how irreconcilable this conflict was may be judged from the fact that many of the issues discussed in Hare's *Miscellaneous Pamphlets* (1855) were similar to those addressed by Rose from an alternative viewpoint twenty-seven years earlier.[43]

Even so, the personal element in all this must not be minimized. After waiting a few months, a hurt and badly bloodied Thirlwall complained that Whewell had twisted his meaning, magnified their differences, and thereby "rendered an immense service to my adversaries: one which they could have received from no other person." Furthermore, by issuing pamphlets "calculated to produce the impression that I am a covert enemy to the Church, who is willing to sacrifice its independence and its highest interests to those of the Dissenters . . . you have . . . abandoned the neutral ground on which you professed to stand, and the friendship which you avowed for me."[44] The injured tone is understandable for undoubtedly, as Jones put it bluntly, Whewell's pamphlets were in effect a defense of the master's demotion of Thirlwall.[45] A more troubling question is: Did Whewell, in part at least, sacrifice an old and intimate friendship for the glory of becoming the next master of Trinity? Probably not, but the possibility remains.

THE DEPARTURE OF HARE AND THIRLWALL

The departure of Hare and Thirlwall from Trinity left Whewell a much more solitary figure. In the shocked aftermath of Thirlwall's demotion, there were fears that sympathetic students might react with violence, and the apostles even considered disbanding in protest. Outside Cambridge liberals celebrated Thirlwall's stand, with Mill describing him as "one of the few great scholars who have issued from either University for a century."[46] Thirlwall swiftly departed for a lucrative living at Kirby Underdale in Yorkshire. Soon after that he was elected to the ungodly senate of London University, an appointment which will have confirmed Whewell's suspicions and made him feel more comfortable about his own part in the events. Their relationship never truly recovered—indeed Thirlwall would never return to the college until three years after Whewell's death. Then in 1840, after fierce lobbying by Thirlwall's Trinity friend, the Harrow-schooled Charles Buller, the Trinity graduate Prime Minister Melbourne offered him the bishopric of St. David's. This was an appointment that brought yet more controversy, particularly among Tractarians but also among many in Wales, which was a hotbed of nonconformity, especially in areas that used the vernacular.[47] Under the premiership of Lord Melbourne, four of the bishops he appointed were from Trinity. He greatly disliked Oxford Tractarianism and ten of the thirteen bishops now all came from Cambridge.[48]

Meanwhile, Hare had earlier departed Trinity to take up the family clerical living of Herstmonceux, Sussex, in 1832. For a time Bunsen became his neighbor and no doubt helped Hare continue his quest for a refashioned National Church that could embody the whole range of theological beliefs and prepare the way for the Kingdom of Christ.[49] As soon as Hare departed, Whewell wasted no time in ensuring he moved into his old friend's impressive college rooms that overlooked the beautiful Lime Tree Avenue to the rear of the college.[50]

"LET FACTS SPEAK FOR THEMSELVES!": THE SOCIETY FOR THE PREVENTION OF CRUELTY TO UNDERGRADUATES

The quarrel over compulsory chapel had divided fellows, and shortly afterward it was to cause them much embarrassment. Feeling that he must expunge any malign influences that Thirwall had exercised over students, Wordsworth published an edict to the effect that attendance was compulsory. Students must attend eight times a week and twice on Sunday. His high-handedness led in 1837 to the creation of The Society for the Prevention of Cruelty to Undergraduates (SPCU), the aim of which was to expose the hypocrisy behind Wordsworth's edict. In February 1838 SPCU began to publish the chapel attendance records of the master and fellows. Whewell, it transpired, attended only twice during the week ending February 17, while several others had not attended at all. The students were ecstatic: "Does, then, this new Regulation of the 'Master and Seniors' proceed from any religious motive? Do they practice what they force on the Undergraduates? They are very regular in their attendance in Hall, but why are their places vacant in chapel? LET FACTS SPEAK FOR THEMSELVES!" A week later SPCU was pleased to announce a slight improvement in the fellows' attendance but warned that they would "not relax in their endeavours to promote the cause of Religion." By now Whewell was up to ten attendances, at which point SPCU announced that it would award a prize to the fellow who exhibited the greatest regularity and good behavior. However, in order to ensure that it was not merely a superficial outward display of religiosity, the winning fellow would have to take an end of term exam. In the following week the society gleefully reported a downfall in attendance and attributed it to too much alcoholic indulgence.[51]

Probably as a result Wordsworth reduced the requirement for compulsory attendance to six hours, at which point SPCU, in one of its final acts, awarded Perry the prize medal—a Bible for regular chapel attendance and good conduct. In addition, other venomous broadsides were published against donnish hypocrisy. For example, in "A New Song to an Auld Fling," the master and fellows were all described in satirical and disrespectful verse. Perry was compared to the Reverend Mr. Stiggens from Dickens's recently published *The Pickwick Papers* (1837); Stiggens was an unpleasant, alcoholic evangelical, hell-bent on material gain. In another work attacking the fellows, "Stanzas," Whewell was portrayed as a "shapeless ogre" with no social breeding—"The coarsest of all boorish clownes, That ever have worn Fellow's gowns"—and having "rambling thoughts."[52]

Undaunted, Whewell throughout this period continued to focus his intellectual attacks on the growing reformist ideology. Something had to be done to prevent the further erosion of the Church of England's monopoly on morality, in particular political economy. This was where Jones was so important to the "true faith" and is the subject of the following chapter.

CHAPTER FIVE

AN ETHICAL ECONOMY

> It seems to be an absurd and humiliating result of so many years spent by Englishmen in the speculations of political economy, that the governing body should carry on the business of the nation on the basis of fallacies so very palpable as those of "Free Trade," and should persist in their course in spite of that diminished national production which, according to the arguments both of advocates and opponents of the policy, must be, as it is, the result.
>
> WHEWELL TO JONES, OCTOBER 29, 1851

Despite Anglican religious doctrinal differences, the TC shared a wide set of views, including a strong dislike of Ricardian political economy. Although Jones was not a Trinity graduate, he was extremely close to the TC and especially Whewell; he was undoubtedly the economic voice of the group. Jones, for example, drew on history to refute Ricardo's deductive economic axioms—especially his theory of rent as something not timeless but a source of revenue that varied from place and period.[1] Ricardo's pseudoscience, with its powerful prophets such as Mill, Nassau William Senior, and John Ramsay McCulloch, was undermining traditional social authority.[2]

The nation and state life was organic, a product of past and present generations. Economic liberalism and the emphasis on the individual were false and politically dangerous.[3] There were no timeless economic axioms; instead things were confined to geography and temporal context. There were no universal economic laws, and the discipline had no fundamental ideas. In his preface to the edited celebration of Jones's work, Whewell reminded readers that it was precisely the defunct deductive political economy of people like Ricardo that compelled Jones to produce a true inductive ethical economy. This is what made Jones so important to the TC. Whewell wrote in a preface, "Many of the conclusions to which he was thus led were, as it may well be supposed, entirely at variance with the doctrines of the *a priori* school."[4] Ricardo had simply produced an ingenious theory based on "purely hypothetical truths," which had absolutely no empirical and historical grounding. Indeed, Jones wrote, "a single comprehensive glance at the world as it actually exists, is sufficient to shew [Ricardo's theory] to be utterly inconsistent with the past and present condition of mankind."[5] Historical induction and not abstract atemporal deduction was needed. Most of Jones's economic work was designed around refuting Ricardo's deductions on themes such as value (labor), distribution, profit, wages, and especially rent.

Since the early 1820s Whewell had constantly prodded Jones to refute abstract Ricardian political economy: "I hope to find the demolition of the Ricardites very forward if I see you, for it is a proper adventure for you to set out to kill such a dragon as that system."[6] Indeed, he hoped they would "partly form a school which would be considerable in influence of the best kind."[7] However, the task was an uphill one. Jones confessed that the "palpably, mischievous & immoral" Ricardian system had "gained upon the public mind the difficulty of gaining a hearing for principles of a different complexion & got at in a different . . . manner & in the spirit of a less presumptuous philosophy."[8] Whewell hoped Jones would be appointed professor of political economy at Cambridge and was dismayed when Pryme got the position: "I do not think it will do the science any good for he is, as a lecturer, dull & unpopular & will never be otherwise."[9] Moreover, he was yet another annoying Whig disrupting ordained social relations and institutions. Even worse, he lurked behind the walls of Trinity.

For Whewell and the TC, Jones had literally been put on earth to slay the abstract doctrines of political economy. This chapter will concentrate on Jones's argument over rent and its relationship to the Church of England. Jones opposed Ricardo's notion of rent not just via history but also the premise that the increase of farmers' rents was always accompanied by a decrease in agricultural productivity. This was a politically loaded view that had dire implications, since it placed the interests of landlords—the Established Church and Trinity being very large ones—as always and necessarily opposed to the interests of every other class. As such, Jones pleaded that once people consulted "the statistical map of the world" based on actual experience, as opposed to the "speculation of individuals," it would demonstrate this was not the case: "We shall see first rising up before us in all parts of the globe this prominent and unquestionable fact;—that under no form or modification of the relations between the proprietors and cultivators are the permanent interests of the landlords opposed to those of the community at large."[10]

Deductive political economy, Whewell hissed, was the product of "the English and French democratic states," and was designed "to preach a theory of the progress of man as a new religion."[11] What he meant, of course, was that this was a human-centric materialistic ideology, not a religion, hell-bent on tearing up traditional social authority. Whewell told Jones to be patient after the latter began to get annoyed by his lack of recognition: Give "the common herd of readers" time to digest your labor—"I have always told you you shall have fame enough, if you will wait a little."[12] However, Jones was impatient and would never quite obtain the recognition he, or Whewell, so craved. The Ricardian dragon was now far too large and had penetrated deep within the country's sinews.

The increasing anger generated by the Ricardian Anti-Corn Law League was, for Jones, fueling the question of rent. It was creating class warfare without any justification. This claim was entirely predictable, since at the heart of his "ethical economy" was a defense of the landed class. The league was created in 1838 by the Ricardian free-traders Richard Cobden and John Bright as a pressure group to repeal the Corn Laws on imported grain; these laws were aimed to protect and keep domestically produced grain high in price (and therefore rent). The Established Church had become particularly embroiled in land ownership during the heated debate over Church tithes—with Jones at the center of their defense. Whewell begged Jones to get his project published: "That you have got to trouble the shortwitted, rotten, pseudo-political-economists; and to yoke history, & morals, & natural characteristics, and practical experience to that chariot of science which they have hitherto been driving tandem with one jack-ass before another. If you think of visiting far lands it must only be to bring back that evidence political Economy which they have condemned to the murky caverns of the nether world."[13] Historical and geographical induction demonstrated that the new political economy was false. Jones's ethical economy would go on to complement Whewell's system of morality.

Land value and agricultural output were an intrinsic component to the Church of England's income and implicit in the question of rent. Rose published a combative tract on the need to defend tithes in *The Farmers and the Clergy: Six Letters to the Farmers of England, on Tithes and Church Property* (1831). Jones too wrote an anonymous article on tithes in Rose's newly established *British Magazine* in 1832.[14] Whewell thought Jones's essay "excellent—if the church conservatives have the sense to take hold of it, there is good foundation there for a popular cry against the innovators—'Here are the rascals who want to take the support of the clergy off the shoulders of the land lords and to throw it on those of the people.'"[15] The innovators being, of course, numerous Whigs, political radicals, a mass of everyday folk, and Dissenters who wanted to scrap tithes and crush the Established Church.

Jones's former Caius acquaintance and large Lincolnshire landowner, Bromhead, was also impressed by his observations. He wrote to Whewell, "[Jones's] idea of labouring classes gradually coming under the domain of capitalists is striking & true." He concluded, the "moral machinery" of industrialization "has not kept pace with the population."[16] For the TC this was because industrialization had helped create a false and irreligious calculating-utilitarian morality. Jones's tactic was to try to deflect working-class grievance away from the Church and landed class to owners of production. No wonder Karl Marx admired so much of Jones's work. Charles Townsend (presumably the Oxford graduate and prominent legal authority William Charles Townsend) also liked Jones's essay and told Whewell it was "full of wisdom & Temperance." Moreover, he concluded, "The Truth is, that the want of us people is really a high, religious & Judaic feeling & belief that the Tythe is an offering to the Lord."[17]

Jones taught his students during the late 1830s about the relationship between the means of production and its impact on the lived experience: "As communities change their process of production, they necessarily change their habits too. During their progress in advance, all the different classes of the community find that they are connected with other classes by new relations, are assuming new positions, and are surrounded by new moral and social dangers, and new conditions of social and political excellence." Jones made it abundantly clear in his work that capitalists were now in charge of the means of production.[18] Marx, with a different end in view, agreed. The former saw it as the eradication of the paternal natural order and, the latter as the exploitation of the working class. This twin axis continued to inform interpretations, as shown in this book's epilogue, as the debate moved into other contested territories.

Although Whewell hailed Jones as the founder of an inductive-ethical economy, the only other leading enthusiast for his work at this time, outside the Anglican Church and among large landowners, was Marx. He greatly admired Jones's historical and non-static emphasis on, for example, work practices and land rents. For Marx, like Jones, changes in production over time was key to understanding the socioeconomic structure and thus contingent on the historical context. This refuted Ricardian universalism and atemporal political economy with adverse implications for the classical theorist's emphasis on profits and wages being closely dependent on the theory of rent.[19] Instead, they depended on the "the laws and customs of the country, by constitution, history and character, so that he cannot, or can only with great difficulty, change his plan and mode of life."[20]

For Ricardo land had a single purpose and that was to grow corn. High grain prices kept less fertile land in production and sustained high rents. It was in the interest of landowners to keep grain prices high. By contrast, Jones claimed, high rents in Britain were a product of improved production and increased capital investment.[21] He argued that those countries throughout time that had very productive agriculture and low grain prices also actually had the highest rents—the reverse of Ricardo's theory of rent. They were also very different. "Serf rents" paid by the cultivator to the landlord in Russia, for example, were based on labor costs; "Metayer rents" in Central Europe were based on produce paid to both the cultivator and landlord. "Ryot rents" in India were paid to the sovereign by the landlord based on part of the produce. Other examples included "cottier rents," a money-rent based on the subsistence of the cultivator, and "farmers' rents"—a covenanted rent paid by a person employing laborers.[22]

Whewell was furious with McCulloch's adverse review of Jones's work in the *Edinburgh Review* of 1831. McCulloch described Jones's work as "extensive" but "superficial," saying absolutely nothing that was not already known.[23] The *Review* was the bane of the Anglican alliance between the Church and ancient universities. McCulloch, a Scottish Ricardian, had totally distorted and deliberately undermined Jones's stance. His friend's point, Whewell wrote, was to demonstrate that rent was not universal and, on the contrary, varied from country to country and over time. You could not construct a system "downwards from two or three simple principles" as Ricardo and his disciples had.[24] Such axioms were false and dangerous.

Jones's first victory came in January 1833, when he was appointed professor of political economy at King's College after the Ricardian Senior resigned. Rose's *British Magazine*, was ecstatic over the appointment. There was no doubt, the Anglican weekly rejoiced, that Jones

> will rescue the science which he professes from the bad hands into which it has often fallen, and the bad character under which it has in consequence laboured. Instead of setting out with declaring certain principles, founded on very narrow observation, to be universally true, and then building up mighty and monstrous theories upon them, he bids the student search, examine, and take comprehensive views of the facts, in order to get at principles, which are truly comprehensive.[25]

Privately Whewell told Jones two years later that he felt the Whigs would destroy the country. If the economy was under the thumb of a mechanistic political economy, it was also under siege by Irish fanatics. Irish radicals in Parliament had teamed up with Whigs to reform the Established Church and, moreover, in a movement led by Daniel O'Connell, repeal the 1800 Act of Union and allow Catholic participation in Irish politics. Whewell was fuming: "It is a thing that rather weighs on one's spirits to know that we are living in a country running rapidly to disorganisation and comparative barbarism."[26] Whewell despaired that "a pride of Irish Barbarians" would conquer England: "It seems absurd to speak seriously of such a danger; but where is the impossibility that the Irish members by being the majority, may carry measures which will brutalize England to the condition of Ireland and worse?"[27] The idea of such a House of Commons meddling in the doctrines of the Church made Anglicans across the country sick with worry.[28] Whewell's gloomy reflections were reinforced in a letter he received from the Westminster- and Oxford Christ Church–educated geologist William Daniel Conybeare, who wrote he could "hardly find words to express the disgust & contempt with which the coalition of that poor tool & fool Johnny Russell [Lord John Russell, Prime Minister] with the Radicals & Repealers gives me."[29] Within this context the career of Jones took on added importance to the TC.

Jones's career seemed to continue to do well during this period, and he subsequently succeeded Malthus to the chair of political economy and history at the East India College, Haileybury, in 1835. Despite his opposition to Malthus's population principle, he shared his predecessor's inductive approach and emphasis on history, morality, and the social structure.[30] The following year the government officially substituted money payments for payments in kind in the 1836 Commutation Act. The archbishop of Canterbury appointed Jones as one of three tithe commissioners to manage the process; Jones's attention now shifted from a global statistical map of land rents to mapping the value of British land.[31]

The tithe was originally a tax on one tenth of the land's produce and had traditionally served as a uniting toll among the land, people, and clergy. By this period, however, it had in many cases been substituted by annual money revenues rather than the value of the harvest or payments made in produce. It relied on a valuation of the land and constituted the predominant income for clergymen. Thus the maps Jones and his two colleagues were commissioned to create were crucial in calculating the value of land. This "corn rent" was initially based on a seven-yearly valuation of the various grain grown on the land.

Many Whigs, radicals, and Dissenters called for the end of tithes—driven by the regular depictions of the clergy as corrupt, lazy, and pointless. The Church of England desperately sought to legitimate its existence and defend its employees. Indeed, the issue had become incredibly sensitive by this point, with many seeing the clergy as, more than ever, the right arm of the landed ranks and as a key element in class demarcation.[32] Cobden was getting very frustrated and wrote, "The Chartists don't seem to understand the real position. They direct all their attacks against capital, machinery, manufactures and trade, which are the only materials of democracy, but they never assail the feudal aristocracy and the state Church which are the materials of the oligarchical despotism under which they are suffering. Fergus [O'Connor] and his demonical followers seem bent on destroying manufacturers in order to restore the age of gothic feudalism."[33] Jones became the adviser to the recently appointed counsel to the Home Office, John Elliot Drinkwater, to help him draft four bills on Church matters—including the 1836 Tithe Commutation Act.[34] Drinkwater was educated at Westminster and Trinity, finishing as a wrangler in 1823. Jones's appointment as one of the commissioners of tithes in 1836 had also been a position Rose had applied for. Many of the maps were first completed by 1841 and the rest finished by 1851.

Whewell confessed to his old friend Rose that Jones's appointment to the Tithe Commission was dashed by Rose's failure to land the post. He told Rose it was high time "that the Church should do something soon if she intends to do it at all—for persons of your standing and mine who have been of any use to her, for we are no longer young men. I hope your turn will come before long, for I have yet to make out my case by *repairing the philosophy of the age*; which I am going to set about in reality" (my italics).[35] Whewell's attempt to soothe Rose's jealousy of Jones's employment proved to be of little comfort.[36] Two years later Rose died.

Outside of the Anglican Church, Jones's emphasis on historical and comparative induction on economic issues made little headway. However, he is the bridge to later English historical economists such as William Cunningham (Trinity), William J. Ashley (Balliol), and Ephrain Lipson (Trinity). Whewell's own aim, which he had already commenced, was to produce both a natural and moral philosophy appropriate to the Established Church and an Anglican university education. In this capacity his own career was about to take a significant and life-changing turn.

CHAPTER SIX

THE NEW MASTER OF TRINITY

By 1840 Whewell was so frustrated with college life that he considered becoming a rural parson; an opportunity at Masham in Yorkshire had arisen, and he asked Hare for advice. He felt his "task to college is well-nigh ended," and his friends were all departing or becoming distant through change of habits. Whewell dreaded "the idea of growing old in college." However, he did not want the life of a country clergyman getting in the way of his writing. Moreover, he admitted to not liking common folk. Whewell was depressed and confused over what direction to take next. Hare was firm and gave him clear advice. He agreed that Whewell was not cut out for life as a clergyman: "In talking to the poor about the petty concerns of their daily life, your life would be a very unsatisfactory one." Hare was resolute that Whewell should devote his life to their college, university, and "repairing the philosophy of the age": "Your ministry in this world seems to me to be that of a doctor, rather than a pastor. . . . I would have you pursue your work in moral philosophy. . . . it would be an evil thing for England that you should abandon it."[1] Whewell did exactly that and concentrated on grounding an appropriate Anglican morality to counter the growing materialistic utilitarian morality. He was further institutionalized into his consecrated role the following year on becoming master of Trinity.

The former master of Trinity, Wordsworth, had deliberately waited for the fall of the Whig government and the election of Robert Peel's Tory government in 1841 before resigning to ensure Whewell was his replacement.[2] Peel was also lobbied in favor of Whewell's credentials by his ecclesiastical adviser, the former Trinity man and bishop of London Blomfield. He wrote to Peel:

> Mr Whewell is a man of extraordinary powers of mind, and his Bridgewater treatise evinces *the soundness* of his general religious views. I have some doubt whether he possesses a sufficient staple of divinity to qualify him for

> the difficult task which I foresee he will have to undertake, of combating or checking the extravagances of the Oxford Tractarians which are beginning to make themselves felt at Cambridge. If he should be appointed to the Mastership *I hope he will feel the necessity of devoting himself somewhat less to science and more to theology.* He will easily master any subject to which he applies his mind. (my italics)[3]

Whewell, unlike Blomfield, was by this stage only too aware that making such a demarcation between science and theology was not so simple. Under the Whig administration it was likely that someone less conservative, such as Peacock, would have been appointed. Indeed, Peacock admitted as much to Whewell: "If the Whigs had been in power I should certainly have been a candidate for a situation which I have always regarded as the summit of my ambition."[4] By this time he had been appointed dean of Ely. There was not much love between these two Trinity colleagues. For Whewell, Peacock erred too much toward the thinking of adverse men such as Babbage in his emphasis on symbolical algebra and justification, via natural history, for the use of negatives and imaginaries. This also manifested in his reformist views toward the university.[5]

As the new master (1841–1866) and, a year later, university vice-chancellor (1842 and 1855), Whewell's power was consolidated, and he could spread his political influence far wider and deeper within the university and beyond. From the moment Whewell was appointed master, he made it clear to Hare what he intended: "It is our duty to conform our whole character, affections, desires, will to the Divine rule, so that we have no promptings to sin but such as we master immediately."[6] This meant, first, all Trinity students and fellows conforming to Whewell's will and preparing for the eternal future. This again was an aspect of Kant's view that sensory experience should not muddy the pure intellect.

In 1844 Whewell revised the college statutes and desperately tried to reduce the emphasis on private tuition and place a greater focus on university lectures. Three years earlier he had told Hare there was "too much emulation and too much private tuition," the "mischief" of which Hare too felt "deeply."[7] While in a letter to a nonresident fellow, Thomas Stephen Gossett, Whewell expressed his approach to college statutes: "My own view is that the changes made should be the smallest which will truly answer the purpose of bringing about an accordance between our laws and our practices; and that we ought to preserve the existing statutes whenever we can without manifest inconvenience." In private Whewell did not approve of nonresident fellows, such as Gossett, being elected seniors in the college.[8] The following year Whewell, in quite a coup, convinced the seniors to add his notion of moral philosophy to the college exams at the end of the Michaelmas term.

Two years later Whewell's brief affair with Peel was over. The prime minister's administration was becoming sympathetic to repealing the Corn Laws and any warmth toward him by Whewell and Jones had evaporated. This had been brewing since at least 1839 when some undergraduates had taken out their frustration about the issue on the Anglican wit Sidney Smith when he came in 1839 to lecture against the Corn Laws in Barnwell.[9] Whewell and Jones were dismayed at Peel's policy toward Ireland. Famine had hit Ireland as the country's crop of potatoes, the Irish staple diet, failed in autumn 1845 and again in the following autumn. Using the potato as the trigger to reform the Corn Laws was, they thought, scandalous. It was also, Whewell thought, a prelude to destroying the Irish Church. He growled, "The sooner Peel is out the better. He has neither political wisdom nor good faith." Indeed, he concluded, Arnold's "judgement of Peel, that he cared about nothing really but finance, is obviously quite correct."[10] In truth, however, Peel had not been motivated to reform the Corn Laws by the potato famine. In fact, he claimed the opposite and that his one reservation was that repeal could destroy Ireland. Moreover, claiming Peel repealed the Corn Laws due to famine is an anachronism, since it was not until the second crop failure that famine hit Ireland, and by then Lord John Russell was in power.[11]

Meanwhile, fighting seemed to be breaking out all over Cambridge. On March 6, during an exhibition in the town hall of the American dwarf Tom Thumb, the police and university students clashed. This was followed over the next few days by fighting between town and gown in Rose Crescent, Market Hill, Petty Cury, and other areas of town. Windows were smashed at several colleges, and many new special constables were recruited. A notice was also published by the vice-chancellor and master of St. Catherine's, Henry Philpott, and all the heads of colleges that anyone involved would be severely punished.[12] Whewell also managed to fall out with Sedgwick that year after formerly writing to him to stop appearing in the college courts with his dog. This, he scolded him, was against the college statutes.[13]

The following year Whewell's temper was once again getting worse, and a couple of days later he furiously told the Trinity porter to stop undergraduates loitering near the gate.[14] Still smoldering in June, he complained to Jones: Who was "going to put out the Fire Robert Peel started in the House of Commons"?[15] A few days later, still moaning, he told Hare they had been seriously let down by those they had previously expected to fight for their views in Parliament: "And laying aside all

PORTRAIT OF PROFESSOR WHEWELL, IN "THE GOLGOTHA," ST. MARY'S.

"Portrait of Whewell," *Illustrated London News*, October 1843. I am grateful to Simon Schaffer for alerting me to this image of Whewell.

strong feelings, I do not see how Peel and his followers can ever again be of service preserving the institutions of the country."[16]

Whewell was as protectionist in religion as he was in agriculture and commerce. Morality was migrating as quickly from Cambridge as it was in Parliament. Whewell groaned to the college tutors that a "mischievous & disreputable" habit of gambling had become established among "some of our pupils." A sophisticated system of betting was taking place in the hall after dinner, with all wagers being officially recorded in books. These practices were strictly forbidden, warned Whewell, and he told the porter "to report the names of those who transgress this Rule."[17]

A reinvigorated Established Church was needed, along with an attempt to instill a related morality. It must be remembered, as the Rugby-schooled and Trinity moral philosopher Henry Sidgwick later noted, it was Whewell who revived and defined the moral sciences at Cambridge—a system that people like Sidgwick and the Merchant Taylors and St. John's–educated Alfred Marshall inherited.[18] It should also be underlined that Whewell's presentation of science was primarily an attempt at making it appropriate for an Anglican university education. This was intrinsic to his objective of laying the foundations of a moral education at Cambridge; it was the link between the theological and material spheres. However, unlike the hard-line approach adopted, particularly by Rose, he still wanted to demonstrate that science need not be the antithesis to subjects such as poetry or theology. Nor was science necessarily a cold utilitarian subject without roots. Indeed, he argued, it evolved with old truths simply assimilated within a new point of view.[19] This was fundamental to his Germanic approach to history and philosophy of the sciences. It was also central to his reputation as a polymath that was often used against him. By contrast, Whewell claimed, this should lie at the heart of a liberal education.

The 1840s were a stressful time for a conservative Anglican educationist like Whewell. At the most extreme end of the spectrum, an "atheist mission" had taken off during this decade, seeking the destruction of Established Church and, indeed, religion in general. For these godless people, there was no innate idea of morality other than that produced via experience of the environment. Thence, they concluded, the obvious thing to do was cleanse the environment rather than the mind. This was evident too in the work of the Benthamite reformer Edwin Chadwick and his work on the new Poor Law and sanitation reform during this era.[20] At the other end of the spectrum a number of High Church Anglicans had and were defecting to Rome, most prominently the leader of the Oxford Movement, John Henry Newman, in November 1845. Thirlwall was dismayed and disgusted.[21]

No institution was safe. Trench warned Whewell that the government was very keen to introduce a commission into the universities with the aim of finding out "whether the universities could not be made adequate to the needs of the present time." The bishop of Oxford had informed Trench that he believed the commission "would contain a good number of sufficiently unfriendly names."[22] Whewell was alarmed. Meanwhile, Hawkins (now master of Oriel) felt it prudent not to be too unfriendly and appear open to the notion of a commission. Now that the Tractarians were no longer so prominent in the university, the institution was far more vulnerable to a liberal agenda.[23]

However, there was one piece of news that cheered Whewell up—Prince Albert was entering the contest to be chancellor of Cambridge University. He told the prince it would be "likely to be of great benefit to the University." What Whewell meant, of course, was that such an appointment would provide great leverage when it came to blocking government reforms of the university's education and rituals.[24] In addition, Albert's Germanic education in Bonn complemented Whewell's philosophical religion, with an emphasis on God as central to the moral world and total obedience to one's conscience.[25] However, before this could happen there had to be an election for the chancellorship between Albert and Edward Herbert, 2nd Earl of Powis, who was being vigorously promoted by his old college, St. John's. Whewell publicly and controversially declared in a letter to the editor of the *Cambridge Advertiser and University Herald* that, if Herbert were elected, he would simply be "a Chancellor of St. John's College."[26] Albert was subsequently elected and became, in all but name, a chancellor of Trinity.

In 1848 Whewell helped to introduce the Moral and Natural Science Triposes. The following year he also offered a bribe of two £15 prizes to be awarded to students who displayed the greatest proficiency in the new exam (increased to £25 in 1851).[27] Meanwhile the clamor for reform of Cambridge and Oxford was getting stronger. One of the leading protagonists was the great geologist, lawyer, Whig, and Unitarian Charles Lyell, who was at the forefront of trying to reform the old universities. Whewell told Hare, "I look upon him as the most bigoted of men in what relates to the Universities." Privately, the feeling was mutual.[28] In terms of the mathematical curriculum Whewell also won an important battle, and geometry was emphasized as the way to understand Newton's *Principia* and not algebraic analysis. Concentrating on the accepted properties of space was, he underlined, crucial to correct pedagogical teaching, and the obscure symbols of algebraic analysis was an evil corruption.[29]

Whewell continued to fervently resist the Royal Commission inquiring into the Cambridge and Oxford curriculum and examination system during the 1850s, believing any reform should be done internally, not externally—in particular, to resist any attempt to shrink the autonomy of the colleges. The idea, for example, of diluting the number of fellowships at Trinity and St. John's to provide a university-wide fund was a detestable thought to Whewell. This would bring him into conflict, once again, with the trio of Sedgwick, Peacock, and Romilly, along with his other associate, Herschel—all of whom sat on the reforming commission. By May 1855 Whewell was bewildered by their actions and the general feeling for reform. He told Everina E. Affleck, "It is a very disgusting and disheartening subject. The bad faith of the ex-commissioners and the democratic frenzy which prevails in the University make it difficult to act or write with any hope of doing good; but I do not think I can help taking up my testimony against what they are doing." Whewell concluded that he felt "disgust, grief, and hopelessness" from "treatment at the hands of friends."[30]

Whewell had very strong views on the university's curriculum. For example, he was fervently against introducing new scientific topics such as chemistry, meteorology, and geology because they were still mainly a mass of observations and speculation, that is, they had no firm fundamental idea linking them together. Only established truths should be taught, otherwise students would become critical, radicalized, and disrespectful of social authority.[31] Nonetheless he was at the forefront of furthering the case for a new natural sciences tripos and moral sciences tripos during the 1850s, and founding the chair in international law after bequeathing money for it in his will; indeed, Whewell, like Kant, connected moral truths with a transnational law.[32]

For Whewell, deluded utilitarian, calculating political reformers seemed to be running rampant seeking radical institutional change everywhere they went. By contrast he wanted institutional preservation through careful, slow, organic change that retained truths from the past. This meant preserving the relationship between the Church and state. As with institutions, so too with science. As he wrote in his *Aphorisms Concerning Ideas, Science, and the Language of Science* in 1840, "When alterations in technical terms become necessary, it is desirable that the new term should contain in its form some memorial of the old one." Likewise, he earlier wrote in his *History of the Inductive Sciences*, "The earlier truths are not expelled but absorbed, not contradicted but extended; and the history of each science, which may thus appear like a succession of revolutions, is, in reality, a series of developments."[33]

Science and society did not progress through revolutions but via slow organic growth. The key to accessing the way human understanding operated was by the study of language and by tracing it as far back as possible to its divine origin. For the ancient Greeks, Whewell argued, "language was an original language—at least in this sense, that it was derived from, not founded upon, any which had been cultivated, enriched . . . diffused and transmuted. The history of every Greek word consisted of only one or two steps, by which it came from some root, modified only by simple laws of thought and speech."[34] The closer one got to original languages the closer one got to God's original given language. No wonder, as we have seen, that Whewell emphasized the study of classical languages along with ancient Euclidian geometry as the basis for a liberal education. Geometry was another original divine language. Whewell told Hare, "A man who either cannot or will not understand Euclid, is a man whom we lose nothing by not keeping among us" in college. A Cambridge education should mean teaching poetry, philology, and ancient Greek philosophy, as well as Euclidian geometry. As it was, the march of the Cambridge factory coaching system was producing uneducated men.[35]

Airy was the perfect example of a proper Anglican education: "He was driven to read Greek by our Trinity system, is very fond of it, and so is an educated man. Without this, he would have been a mere mathematician; and such a one is not an educated man."[36] Whewell explained precisely what he meant by the importance of tracing the history of language:

> The words in modern language had their origin in Greek or Latin, but they have had the meaning of their roots, or the significance of their derivatives, modified by the thoughts, the habits, the accidents, of Greek and Roman cities; by currents of population, determined by war or migration; by the currents of the thought of the middle-ages; by the further progress of thought and speculation in modern times; while, in the course of these changes, the structures of ancient languages have been broken up, and recast, and rewrought by the processes of *new grammatical machinery*.[37]

It was in this sense that the application of German hermeneutics clearly informed Whewell's approach. It was only through the close archaeological scrutiny of language that one unraveled the corruptions of language and eventually the underpinnings of the original divine language: "On the structure of language are impressed the most distant and durable records of the habitual operation of the human powers. In the full possession of language, each man has a vast, almost inexhaustible, treasure of examples of the most subtle and varied processes of human thought."[38] German hermeneutics was a tool to work back to the purity of God's original language. However, by now Whewell differed from Hare and Thirlwall. It was not a tool to undermine the gospels but to truly understand them; not to question the apostles but to be as one with them. Despite their differences there was now no difference between Rose and Whewell's religious outlook.

CHAPTER SEVEN

GENIUS AND GREAT SCIENTIFIC DISCOVERIES

> A *formula* is a brief mode of denoting a rule of calculating in which numbers are to be used: and numerical measures are applicable only to things of which relations depend on time and space. By such elements, in such a mode, how are we to estimate happiness and virtue, thought and will? To speak of a formula with regard to such things, would be to assume that their laws must needs take the shape of those laws of the material world which our intellect most fully comprehends. A more absurd and unphilosophical assumption we can hardly imagine.
>
> WHEWELL, 1833

Whewell and the TC moved the emphasis of morality from Paley's sensory-mechanistic-external design argument to the pious and inner character of great natural philosophers. The analogy between man-made devices and the natural world was not just redundant, but was complementary to a materialistic and deist perspective that characterized prevailing political radicalism, wrong-headed Dissenters, and pragmatic Whigs. "Genius," argued Coleridge, was "itself an unconscious activity; nay, that is the genius of the man of genius."[1] Kant defined it as "the inborn mental trait (*ingenium*) through which nature gives the rule to art."[2] Where Newtonian natural philosophy found natural theology in the heavens, Paley, like Babbage, found it in terrestrial industrial machinery, with the human body (including the brain) being nothing more than a mechanical assemblage.[3] For many this meant God was simply the architect of matter but no longer actively intervening in his creation.[4] This may have managed to bridge and appeal to a whole gamut of religious factions, but for the TC it was just plain wrong.[5] Vitally too it collectivized the process of discovery—something closely associated with the French Revolution and potentially a tyrannical democracy. Discovery was something far different and important than a collaborative and mechanical process of the mind.[6] This also underpinned Whewell's and Hare's hostility toward the industrialization of the Cambridge Tripos—a point we return to in a later chapter.

Hare taught that those "who are to attain to any eminence in knowledge, must have a strong Faith in the desirableness of knowledge for its own sake," and lamented that Cambridge was neglecting this divine element within its curriculum. The reason modern Europe had diverged from the rest of the world in science was "owing wholly to the influence of Christianity," but it was also the same Christian faith that prevented us from descending into empty materialism.[7] This last aspect was fast being eclipsed and thus so too was true knowledge and morality. No wonder Whewell devoted the rest of his life to trying to define a universal morality, ultimately derived from divine intuition manifesting in certain ethical laws and contingent duties. This again was informed by a particular interpretation of the work of Kant. Whewell's first public attempt at trying to establish a divine-based human morality was probably in his Bridgewater Treatise.[8]

Euclidian geometry revealed certain divine ideas. This, as Peter Harrison has shown, had a long history. Tradition had it that at the entrance of Plato's Academy were the words "Let no-one ignorant of geometry enter here."[9] Thomas Aquinas claimed geometry taught men the correct way of thinking.[10] Much later Galileo Galilei wrote in his *Dialogue Concerning the Two Chief World Systems* (1632), "The human intellect does understand some of them [mathematical truths] perfectly, and thus in these it has absolute certainty as Nature itself has. Of such are the mathematical sciences alone; that is geometry and arithmetic, in which the Divine intellect indeed knows infinitely more propositions, since it knows all. But with regard to those few which the human intellect does understand, I believe that its knowledge equals the Divine in objective certainty, for here it succeeds in understanding necessary, beyond which there can be no greater certainty."[11] Galileo concluded that the book of nature was "written in the language of mathematics, and its characters are triangles, circles, and other geometrical figures without which it is humanly impossible to understand a single word of it."[12]

Similarly, and slightly earlier, the leading sixteenth-century Protestant scholar and associate of Martin Luther, Philipp Melanchthon, held the view that through the study of mathematical astronomy of the uncorrupted heavens, we could find the lost divine moral order on earth. Harrison writes, "In a sense, Kant's later coupling of 'the starry skies above' and 'the moral law within' had precedents in the thought of Melanchthon."[13] God was a geometer, a view which impregnated pious men such as Kepler and Newton. Kepler was a follower of Melanchthon and wrote in his *Harmony of the World* (1619), "Then since they [spirits, souls and minds] have embraced a certain pattern of the creation in their functions, they also observe the same laws along with the Creator in their operations, having derived them from Geometry."[14] No wonder Whewell worked to place his system of morality on the same level as geometry. His concern was a medieval one, namely to nurture the spiritual and moral and not create new knowledge. Moreover, insofar as he was concerned in the latter, his work on the philosophy and history of the inductive sciences was to demonstrate that such a spiritual morality was intrinsic to all great scientific discoveries. The whole purpose of a liberal education for Whewell was the medieval quest to restore some image of the Divine as it was prior to the Fall.[15]

A controversy that revealed Whewell's emphasis on a distinctive morality as intrinsic to the scientific process erupted in the mid-1830s. Whewell was incensed when the retired stockbroker and actuary Francis Baily, also a leading member of the Astronomical Society, attempted to rescue the importance of John Flamsteed's role in Newton's scientific discoveries. If it had not been for the first astronomer royal's skilled astronomical observations, argued the Unitarian, Newton could never have come up with his gravitational astronomy. This was the same argument, as we will see, that Mill made about Kepler's detection of elliptical orbits. Sensory observations of Flamsteed had revealed to Newton the pattern of gravitational astronomy. As Flamsteed himself had underlined to Newton: "The works of the Eternal Providence will be better understood through your labours *and mine*" (my italics).[16]

A primary message of Baily's reinterpretation, published in 1835, was that beneath scientific discovery and progress was a great deal of careful, laborious, and precise skilled work. Flamsteed provided the model example of a systematic, conscientious, and disciplined practical astronomer. His astronomical accountancy was of a nature hitherto unsurpassed. Crucially, he carefully retained a record of all his original computations. This was, for Baily and the leading members of the Astronomical Society, fundamental to the value of his observations.[17] This is an argument that has carried on to the present day.[18] It enabled contemporary astronomers "to detect many errors in that catalogue [the *British Catalogue*], but also to discover the source of them, and thus correct them with more confidence."[19] If genius and guessing at truth were to be regarded as the vital part of science, then it would undermine the primacy of regular, mechanical, rule-bound observation—the kind of science emphasized in London and especially at the Astronomical Society. There would be no lone imaginative genius without the labor of the calculating clerics.[20] Moreover, beneath this argument was a reinforcement of Locke and Hume, namely our knowledge derived through the senses and not innate ideas.

The former Trinity student, mathematician, actuary, and leading member of the Astronomical Society, Augustus De Morgan, similarly criticized Newton for having tended to treat his results as private knowledge and failing to give credit to the importance of Flamsteed's observations. Secrecy was a threat to scientific verification, knowledge production, and progress. Newton had literally stolen Flamsteed's property, since the observations had been produced at his expense with his own instruments.[21] An outraged Whewell interpreted this as an attack on the morality of a pious Trinity giant and a deliberate materialistic-sensory trend undermining the divine process of scientific discovery. He was also still smoldering and feeling vulnerable from the Thirlwall affair. He wrote to Jones, "You know all Trinity men *must* care for Newton," and De Morgan was thus clearly no true Trinity man—besides, like Baily, he was a Unitarian and materialist sympathizer.[22] He was typical of the sullied metropolitan reformer excised of true faith and deluded by false doctrines based purely on human reason and sensory empiricism. If scientific discovery was merely a human mechanical enterprise it had leveling implications that greatly undermined social authority, organic institutions, and an elite education; Trinity was not some two-penny counting house. Newton was both a philosophical genius and exemplary model of Christian morality.[23]

For Whewell, and the TC in general, the scientific study of God's creation required a necessary moral demeanor, someone with a soul, a pious person like the genius Newton. Hare preached, "Only through Faith, and by that patience and perseverance which a firm Faith alone can give, has knowledge ever been increast and exalted."[24] Those who made great discoveries—such as Copernicus, Kepler, and especially Newton—possessed the religious disposition necessary to reflect, albeit dimly, the mind of the deity.[25] Even here, however, "fundamental ideas" of space, time, force, and matter would still be "far too obscure and limited to be regarded as identical with the Divine Idea."[26] Humans could only ever hope to gain a glimpse of God's ways and purpose—a shadow illuminated by an inner lamp of true faith. The thrust of this perspective was a divinely appropriated version of Kant's synthesis between experience and innate ideas.

Consequently, one of the most damning issues stemming from Baily's onslaught concerned Newton's suspected Unitarianism and Edmund Halley's alleged atheism. How could such un-heavenly minds be so embedded in Anglican education and harnessed in support of moral standards? It is unclear how much Hare and Whewell knew about Newton's actual religious proclivities; for example, Newton privately thought "the central Christian doctrine of the Trinity was a diabolical fraud."[27] Certainly by the time David Brewster published his second history of Newton in 1855, it became clear that Newton had spent a great time researching alchemy and was a staunch anti-trinitarian. However, the view of Newton as a genius remained into the twentieth century. As Rebekah Higgit has shown, Newton got used by opposing groups and their promotion of a particular view of science.[28] Baily mischievously implied that if anyone fit the TC's Christian credentials it was the devout Anglican Flamsteed: "All his letters breathe a spirit of piety and resignation to the will of heaven." It was not "to be expected," he claimed, "that the loose and irreligious conduct of Halley, both in his conversations and in his principles, could be at all congenial to a mind constituted like Flamsteed's."[29] In this reading only Flamsteed had a soul. This humorous, but venomous, sarcasm was designed to hit at the heart of the Established Church, Oxford, Cambridge, and all their sinews. Whewell's vision of devout true science was exploded by how far great discoveries fell short of it, indeed violated it.

The institutional alliance that had grown up between Newtonianism, Christian epistemology, and the Church establishment was on trial. Baily's questioning of Newton's and Halley's theology was carefully crafted to strike a blow at this holy alliance. Newton had held the Lucasian Chair of Mathematical Science at Cambridge, and Halley had been Savilian Professor of Astronomy at Oxford. During the eighteenth century Newton's *Principia* had developed into the staple intellectual diet of Cambridge undergraduates. His work was treated as revealing the plan of Creation, while the plan of salvation was revealed in the sacred texts.[30] Newton had the right demeanor and disposition—he was closer than others to God and could therefore see what others could not.

Stephen Peter Rigaud who, like Halley, held the Savilian Chair of Astronomy at Oxford, was determined to strike back and told Whewell, "If the Flamsteedians will not be quiet, I am ready to fight knee-deep for Newton, if you will enter into an alliance, offensive & defensive with me. Between [us] we win." Rigaud planned to produce a biography of Halley, which according to his son appeared "a natural duty." It would have been "partly in the cause of science, and partly also with the view of rescuing his memory from much of that injurious obloquy to which it had been exposed."[31] Newton, and to a similar extent Halley, were embedded within Oxford and Cambridge. To question their word and moral attributes was literally to question the integrity of these two ancient universities and the holy alliance between the Established Church and the state.[32]

Whewell had already produced a pamphlet in 1836, publicized as a response to the ever-troublesome onetime clerk and now second secretary to the Admiralty, Sir John Barrow, who had recently published a positive review of Baily's book. Whewell's response was clearly designed as an antidote to Baily's specific arguments. His first point was simple: if Flamsteed had been so bitterly abused by Newton, why was he also so disliked by all of England's top men of science, such as Halley, David Gregory, John Arbuthnot, Richard Mead, and Christopher Wren. Conversely, no one other than Flamsteed and William Whiston spoke so disparagingly of Newton. And Whiston's judgment, smirked Whewell, "is perfectly worthless, for he was a prejudiced, passionate, inaccurate and shallow man."[33]

The technical issue of the dispute concerned the hierarchy and mental operations involved in discovery and the question of intellectual property. The latter, as we have seen, concerned the status over the publication of Flamsteed's observations. For Whewell there was no question as to who the guilty party was: "The purpose for which Newton desired that the world should possess the best observations, was the confirmation of the great Theory of Universal Gravitation;—incomparably the greatest discovery ever made by man."[34] Newton's religious disposition gave him access to the fundamental idea of (gravitational) force. Consequently, any dispute over the ownership of Flamsteed's observations simply evaporated into insignificance: the drudgery of mechanical observations (something anybody could do with the right training) versus the greatest theory ever (something only a select few with a spiritual disposition capable of recognizing divine fundamental ideas could achieve). Where was the contest?[35] At stake was Whewell's devout Anglican scientific epistemology.

The dispute was not over the importance of gravitation but the path toward discovery and verification. Whewell acknowledged he was no genius like Newton, but via his devout belief, he could trace Newton's inductive path to gravity. Whewell acknowledged he would never be remembered for a scientific discovery, but he might be for defining the pathway and those that occupied this road; this was the motive behind his history and philosophy of the sciences. Discovery was anything but a mechanical process. Baily, following other founding members of the Astronomical Society, most notably Babbage and Herschel, viewed intellectual production as akin to material production.[36] The detection of gravity was made possible only by the quantity and quality of the raw materials (observations) made by Flamsteed; this also dictated the value of the finished manufactured good, be it material or social truth. The system of production was just the logical workings of the human mind that ground out the knowledge. Herschel told Whewell exactly what he thought of his philosophy in a letter written in August 1840: "You are too a priori [more appropriate would have been metaphysical] rather for me—as soon as one has worked ones way up to a general law you," he scoffed, "come cranking in and tell me it is a fundamental idea innate in everybody's mind—a necessary truth or very likely to prove such in one thousand years hence."[37] Many found this hybrid Kantian aspect of Whewell's metaphysical philosophy abhorrent. For Whewell, fundamental ideas such as space, time, and number underpinned geometry and arithmetic, with force and matter lying at the heart of the mechanical sciences and verticality at the center of architecture.

A few months later, while working on his review of Whewell's *Philosophy of the Inductive Sciences*, Herschel further warned, "I fear talk as we might, or write as we may shall never convert each other. We are like two staunch politicians Tory & Radical who agree in love of country and whom a thousand delightful associations keeps from tearing each other's eyes."[38] Conversely, Whewell and the TC viewed scientific discovery as a much more spiritual process in which the few great scientific advances involved the soul, induction, and imagination. As Whewell later put it, "The Divine Mind must reside in a *Divine Person*."[39] Present scientific knowledge was also connected and inseparable from former generations of knowledge. However, retaining theology as the center of such a process was becoming unacceptable to materialistic political reformers and certain men of science.[40] Whewell had earlier told Herschel, "To begin, after our ancient fashion of disputatiousness, with the points on which we differ, I am inclined to think that you dispose rather too lightly of the ancient languages, as instruments of giving a philosophical civilization to the mind." The study of Greek and Latin literature, Whewell repeated, was "an essential part of the liberal education of the present age"; following Kant, Schelling, and, in general, German Romanticism, it set the standard to gauge beauty—an aesthetic imagination that reveals ontology.[41]

By contrast, and much more in the mold of Herschel and Babbage, De Morgan wrote that the operations of discovery were "more like a book-keeping operation than the poetical process of the fable"—a clear dig at Whewell's pious Anglican and Romantic view of science.[42] This was an irresolvable debate: on the one hand, the accumulation of observations led to discovery and on the other, the discovery via observations only made sense because of the innate idea. For the TC, God created the world and humankind with his design embedded within us (not externally) via fundamental ideas.

What was needed now was someone like the celebrated Trinity philosopher Sir Francis Bacon to do for moral philosophy what he had done for natural philosophy. Certainly Whewell, although far from being a devout Baconian, saw this as his divinely appointed role on earth and devoted the latter half of his life establishing an appropriate moral philosophy. The sensory monists and rational deductivists were creating a world that merely undermined traditional social authority, secured by an amoral utilitarianism, and they had to be crushed. No wonder Whewell commissioned a statue of Bacon to adorn the college's chapel in 1845, followed by one of Isaac Barrow in 1858. A few years later they would be joined by a statue of Whewell himself. Interestingly, one of the remaining three statues was of a fervent follower and student of the TC, namely Alfred Tennyson (1909). The remaining statues are of Thomas Babington Macaulay (1868), a onetime student of the TC but not a follower, and, of course, Isaac Newton (1755).[43] The legacy of the TC may not have won the political and intellectual debate, but their memory is preserved in much of the marble within the college's most hallowed space. Progress for Whewell and the TC was not in the material sense but via the perfectibility of morality derived through true religion.[44]

JOHN STUART MILL AND THE CASE OF JOHANNES KEPLER

A similar debate to the above occurred with Mill. For the TC the present was both a product of and dependent on past historical developments. In short, it was alive, organic, and certainly not mechanical. Whewell tried to show that discovery in science—like the development of British institutions—was dependent on what had preceded it, which in turn could direct future studies. Using Johannes Kepler as his example, he asserted that great scientific discoveries involved imagination and speculation. This flew in the face of Enlightenment approaches to discovery and historiography. Instead, claimed Whewell, truly great men of science were not bound by rules and often moved beyond them. Consequently, he did not seek to ground the rules of science but rather "to generalise the pattern by which these discoveries were translated into scientific disciplines." Science was the accumulation of its past and moved via a set of generalizations and imaginative guesses.[45]

Kepler had discovered the elliptical motion of Mars's orbit only because he possessed the idea of an ellipse; others before him had access to these observations but lacked the necessary divine spirituality to access the idea. The process of induction ended when God revealed it to the mind—that is, a mind with a soul. This had been one of Whewell's central arguments in his Bridgewater Treatise.[46] As he later summarized in his *Philosophy of Discovery*, "The mind performs the operation by an apprehension of some such relations; by singling out the one true relation; by combining the apprehension of the true relation with the facts; by applying to them the Conception of such a relation."[47] This relational aspect was not given in empirical sensations and observations; rather, an inductive discovery was thus a combination of the concrete external data and the mind's divine intuition. It was ultimately the latter that molded and made the connection to what is out there. It was only then it became a "necessary truth"; hitherto it had just been a "contingent truth."[48] As Coleridge had taught, discovery could happen only if you already held the right idea and therefore spiritual temperament; non-divine intuition—purely human reason—led to false knowledge.[49]

Whewell's historical approach put him at loggerheads with prevailing political reformers and especially one of the leading British nineteenth-century philosophers, Mill, and indeed the clerical logician Baden Powell at Oxford. Mill had deep reservations concerning Whewell's theory of intuitive conceptions and fundamental ideas as the basis of a "philosophy of discovery." By contrast, Mill argued that Kepler's discovery of Mars's elliptical motion was not via some innate idea he possessed but rather something found in the sum of the observations. He discovered the conception in the observations that in turn illuminated the ellipse.[50] Babbage and Herschel took the same line. It was the mill of human reason that detected the pattern in the observations and not some innate idea. This, as Joan Richards has shown, was a major epistemological issue: "For Mill, the empiricist, mathematical ideas were somehow constructed inductively from experience; for Whewell, the nativist, they were innate."[51]

In private, Whewell told Jones he found Mill's perspective offensive, on this occasion referring to Mill's *A System of Logic Ratiocinative and Inductive* (1843). This metropolitan theorist was, according to Whewell, "quite subjugated by one whom I think a very bad philosopher, [Auguste] Comte [1798–1857], of whom I could easily show you is a most gross idolatry." Moreover, Mill relied far too much on the new work of Justus von Liebig and Herschel's version of William Charles Wells's work *An Essay on Dew* (1814), in a lengthy engagement with Wells's essay in his *Preliminary Discourse on the Study of Natural Philosophy* (1831); indeed, he scolded Herschel for reducing scientific discovery to this single example. Mill's work was, Whewell concluded, "deadly dull."[52] Jones was equally damning of Mill. He was dismayed a year later over a paper published by the logician "to prove that a priori reasoning is not only good in Polo. Eco. but the only reasoning applicable to it. Lord help him & those this belief leads to trust in him, his Papa & School."[53] Mill acknowledged the crucial importance of inductive observations but claimed human a priori reason was enough to process the data. God was not operative at all in the activity. Mill was not quite a logician in the Richard Whately sense but still believed, like Babbage and Herschel, that human reason alone was enough to decipher sensory observations.

INSTITUTIONAL PRESERVATION AND THE "TRUE FAITH"

A clear battle played out in the clash between Whewell's and Mill's versions of scientific discovery and method. This was explicitly denoted by Mill, who condemned the elitist and conservative intuitionist philosophy of Whewell as simply a device to preserve "false doctrines and bad institutions." Mill wanted his philosophy to eradicate such views and reform British society free from Established Church doctrine and its accompanying philosophy. Experience and human reason, rather than divinely ordained inner feelings or theological prescription, should be the basis for the way things are and should be. The pillar of God, through the Church of England, had to be pulled from the edifice of elite social authority.[54]

By contrast, Whewell sought to mold science to meet the institutional preservation of the Established Church and its centrality to social authority. Material, economic, political, and moral truths had to be addressed in this fierce battle for the mind and legitimation of societal rule. Science evolved via a series of historical developments characterized by divine fundamental ideas that defined knowledge and morality.[55] As such, earlier truths were absorbed and extended in a series of developments. This was a process akin to Coleridge's notion of a plant growing, as opposed to a Babbagian atemporal machine churning out knowledge. Whereas Mill emphasized a logical theory of proof, Whewell underlined an organic divine form of historical discovery; for the former induction was simply an empirical and analytical human process, whereas for the latter it involved an innate idea reflecting God's blueprint to collate the observations together.[56]

In addition, while Mill was both a suspected irreligious sensory empiricist and a priori deductivist, someone like Ricardo was too much to the opposite end of the spectrum and lacked any empirical proof for his argument; both perspectives were equally guilty of ignoring God. Whewell sought to show—unlike Mill, Babbage, and Herschel, for example—that science didn't progress in some simple mechanistic manner, and turned to history to demonstrate this; "fundamental ideas" were independent of experience. Like Mill, Herschel remained hostile to Whewell's metaphysical argument all his life. Indeed, Mill, as we have seen, modeled his philosophy of induction on Herschel's *Preliminary Discourse*. It was human sensations excited by matter that made manifest external objects, and this was done independent of any divine inspiration.[57]

Whewell wrote in his *Mechanical Euclid* (1837)—a book Jones described as the "pocket pistol"[58]—"Some notion is *superinduced* upon the observed facts. In each inductive process, there is some general idea introduced, which is given, not by the phenomena, but by the mind."[59] Bang! Three years later he reiterated his argument: "The conception by which Facts are bound together, are suggested by the sagacity of discoveries. This sagacity cannot be taught. It commonly succeeds by guessing; and this success seems to consist in framing several *tentative hypotheses* and selecting the right one. But a supply of appropriate hypotheses cannot be constructed by rule, not without inventive talent."[60] Thus science, demonstrated by history, revealed a causal structure of nature classified by natural types only discoverable by a philosopher with a soul (someone with "inventive talent"), which was a mirror of God's creation. The physical and moral world were inseparable and made by God. As such the Creator had given the mind the innate seeds of the ideas to understand this world; for these ideas to be revealed required a deep divine faith and correct morality to add the nutrients necessary to grow the idea. This growth also demonstrated the need to show the historical trajectory of such ideas and the right Anglican education.

For the TC in general it was statements such as that pronounced by the nineteenth-century editor of Paley's works, the blunt-spoken, newly appointed archbishop of Dublin and Oriel logician, Richard Whately, in his *Introduction to Political Economy* in 1832, that caused them so much anguish. He claimed the scientific basis of political economy "must be the [mechanistic human] reasoning-process;—the accurate and dexterous application of Logical principles, in combining and drawing inferences from, those few and simple data from which we set out;—in short, the Logical, not Physical investigation." Whately emphasized that there was an objective logic, as in algebra, that could use symbols to substitute for things. This was the basis of correct reasoning.[61] This could have been taken from Herschel's and Babbage's philosophy of analysis espoused during the heyday of the Analytical Society at Cambridge.[62]

Whewell chastised such ungodly deductivists as Ricardo in his review of Jones's *Essay on the Distribution of Wealth* and specifically the support such a deductive approach received from Whately and similarly minded men at Oxford—primarily Oriel.[63] Here Whately distinguished between science and morality and saw nothing un-Christian in focusing on self-interest and a deductive system of political economy. Indeed, as Bernard Mandeville had earlier argued, a single bee was like a single person with both pursuing their individual interests for the common good, in the case of a bee toward the hive and of a human toward society. Man as a rational agent was, for Whately, demonstrative of "divine Wisdom."[64]

Jones thought Whately's first book on logic contained "a strange notion that inference made by the mind when its inductive processes are completed is an inference which belongs to deductive reasoning & may be called logical ascendancy to his own strict definition of logic which I hold to be utterly wrong." He also thought Whately's pupil, the Etonian and Magdalen graduate Nassau William Senior and his work on political economy to be "nonsense".[65] Whewell could not wait to see Jones publish his views on induction, "for it is certainly an important one how the true faith can best be propagated." Whewell had done his utmost to do so in his recent review of Herschel's book.[66] Faith based just on human-defined facts, he stressed, ended up being wrong faith, since the facts took over and never made the last steps bridging the temporal and the eternal world.

Jones assured Whewell he would "help support the faith & have all the disinterested zeal of a knight errant in the great cause." He was so confident, he boasted that he would even come to convert Whately.[67] Predictably, this did not happen. Whewell wrote, "I am quite ready to fight and very confident of getting [the] better" of Whately. However, they needed more historical and global examples, and thus the need for Jones to get his work on wages published: "The feeling on which Whately grounds his opinions, that principles of action are known by consciousness and do not require detailed observation . . . is plainwith and generally assumed and will take some trouble to eradicate."[68] It was only by accumulating examples "of the ascending method applied to moral sciences" that "the downward mad people" would be defeated.[69]

Political economy did not exist—it had no fundamental ideas—and history showed this. It was the personification of corrupt human thinking and needed to be replaced by a clearly expounded morality—an ethical economy. As we will see in the next chapter, Whewell now devoted his life to forging a pedagogical moral philosophy at Trinity. He advised Jones to either "neglect" Whately "or kick him as you like. You will of course soon be as completely out of his reach as a man who walks forwards is of a man who stands still and prances."[70] The opposite, of course, would come to be.

CHAPTER EIGHT

TRINITY AND MORAL PHILOSOPHY

> I hope in particular that the separation of the Morality of Religion from that of mere [human] Reason will be approved of. This separation enables us to trace the results of moral guidance of human Reason consistently and continuously, while we still retain a due sense of superior authority of Religion.
>
> WHEWELL, 1848

> The Ideas of right and wrong—the *moral* Ideas of Man—are then a part of the scheme of the universe to which man belongs. Could they be this, if they were not also a part of the nature of that Divine Mind which constitutes the universe?—it would seem not: the Moral Law of the universe must be a Law felt and discerned by man.
>
> WHEWELL, 1860

A year before the Flamsteed controversy, Whewell felt he was close to a solution for an appropriate pedagogical system of moral philosophy; he had gained "a glimpse, which I have long been wishing and struggling for, of the inductive history of ethics." His primary inspiration was the work not just of Bishop Joseph Butler, but that of the jurist and professor of law and general politics at the East India College, James Mackintosh. It was the later work of the Scotsman that Whewell admired and certainly not his earlier defense of the French Revolution. By this stage Mackintosh had become a favorite among slave abolitionists such as William Wilberforce.[1] Mackintosh vehemently turned against his initial support of the Revolution after the Terror. Ironically, after having opposed Edmund Burke's interpretation, like so many others, he now embraced it. It was Mackintosh's *Dissertation on the Progress of Ethical Philosophy* found in the seventh edition of the *Encyclopaedia Britannica* (1831) that Whewell so admired. It was particularly the anti-Benthamite element in the work that appealed to Whewell—precisely the part

James Stuart Mill had slammed in his "Fragment on Mackintosh" in 1835.[2]

Whewell, with a very lengthy preface, had Mackintosh's essay republished the same year as Mill's review.[3] He lectured to Jones, "That our moral judgements belong to a peculiar *faculty* indeed, and cannot be resolved into judgements, concerning mere [Benthamite] pleasure and pain is Mackintosh's opinion and mine, but is far from new. The arguments which I have given in favour of this opinion are put in my own way rather than his and were intended to be the main nature of the essay."[4] Hallam told Whewell he thought he had "conferred a great benefit on the public by rendering Mackintosh's dissertation so generally acceptable; it will really show that there is a want of interest in ethical philosophy, if it now fails to become a standard book. Your own preface, being a clear analysis of the leading parts of the dissertation, will undoubtedly contribute to the result."[5] Mackintosh's work could appeal elsewhere, but only because it was interpreted differently. For example, Darwin embraced the idea of morality being instinctive but for him devoid of any direct divine interpretation.[6]

By this stage Whewell, who had been feeling very lonely,[7] was thrilled when his old friend Thomas Worsley returned to Cambridge. He had been a contemporary of Whewell, Hare, Thirlwall, and Rose as an undergraduate at Trinity. He had then become a clerical fellow of Downing from 1824 and later a member of the Cambridge Camden Society. He took a brief clerical living elsewhere but returned in 1836 as master of Downing. Along with Whewell, Christopher Wordsworth, French, and others, he became a patron of The Society in Aid of the Incorporated Society for the Propagation of the Gospels in Foreign Parts (1842).

Hare rejoiced when Worsley was made master of Downing: "He will at least make Downing a place of some meaning, in spite of its architecture." Here he was probably referring to its design in the Greek-revival style instead of Gothic.[8] In 1842, Whewell described Worsley as "my especial friend and companion"—not only someone he could talk to about moral philosophy but a riding and holiday companion. They had both reached close agreement by tackling the subject in different ways. Worlsey approached it on the "a priori road, through the views of religious teaching, and advancing to it on the other side, from psychology through jurisprudence."[9] Earlier Worsley, as vice-chancellor in 1838, had probably been instrumental in making Whewell the new professor of moral philosophy at Cambridge that year.[10] Confident with his new ally by his side, Whewell embarked once again on purifying the education at Cambridge. He confirmed to Hare, "I think that the Scripture is good, as proof *against* those who construct a system of morals in which the reference to conscience must be considered as an evasion, which is the case with the utilitarian systems."[11]

Hare was keen that students should be taught to pursue knowledge for its own sake rather than being bribed by prizes. He may not have agreed with Pusey and Newman but admired what they had done for Oxford and wished Cambridge could find such professors of divinity. In addition "the efficacy of the lectures" at his old alma mater "is almost destroyed by the never-ending still-being examinations." Hare concluded that a certain amount of divinity should be compulsory to anyone who passed a degree.[12]

In terms of possible divinity professors to fill the vacant chair at Cambridge in 1841, Hare suggested his brother-in-law and former pupil, Maurice. Whewell agreed and thought Maurice's *Kingdom of Christ* (1838), once introduced, would have a very positive impact on Cambridge students. Maurice's notion of a Coleridgean national church, this time based on history rather than metaphysics, complemented Whewell's use of Coleridge. Niebuhr's *History of Rome* (translated by Hare and Thirlwall) also helped shape Maurice's thinking. Not to be outdone, however, Whewell also reminded Hare that he had already used his own "influence to introduce an anti-Lockean philosophy, and intend to use it for other good purposes."[13] In the meantime Maurice failed to land the position.

By 1845 Hare was relieved that the "theological strength of the University" had greatly improved thanks to the teaching of Worsley and Trench. He told Whewell he was optimistic they would "be good watchdogs against the Oxford hysteroproterites [Greek for putting the cart before the horse]. I am very glad you concur so entirely with my Church-views."[14] Whewell also benefited greatly in writing his *Elements of Morality, Including Polity* (1841) from his friendship with another Trinity man, William Empson. Empson was the successor to Mackintosh at the East India Company College at Hailesbury (1824–1852).[15] In 1841 Whewell published this work on morality. The two volumes were dedicated to William Wordsworth for revealing the essence of morality in his poetry. Even more important, however, was Bishop Joseph Butler for so clearly expounding a morality of such a higher order "than mere tending to produce pleasure," as Paley and Bentham ordained.[16]

Whewell emphasized that the fundamental object of a liberal education was "to develop the whole mental system of man and thus to bring it into consistency with itself." He repeated Hare's, Thirlwall's, and Rose's attacks on the danger of just nurturing the faculty of human reason and relying on the senses: "that there exist vast and solid edifices of truth, the foundations of which are not laid in the information which our external senses give." Crediting truths to an empirical origin was wrong; rather, their roots lay deep in our minds and their dialectic with the material world. This had been the basis of his attack on Baily and, indeed, Mill.[17]

Myers, along with others, was uncomfortable with Whewell's association of morality with geometry in his *Elements of Morality*: "To claim for any elements of morality a foundation philosophically similar to that of the elements of Euclid seems to me so much too much as to permit, even from the most pacific, at least a private protest." Whewell replied by telling Myers they had the same view of morality but a different one of geometry. He had wanted to use mechanics rather than geometry but felt it would have offended too many people. Nonetheless, mechanics illustrated his point better since they now had axioms that had not always been so well defined: "We can point to persons, and to whole ages and nations, who did not assent to those axioms because the mechanical ideas of their minds were not sufficiently unfolded." He then came to his point: "In this we have, as seems to me, an answer to the [Myers's] objection that what I assert as moral axioms are not evident to all men. They are as much evident to all means the axiom that a body will not alter its motion without a cause, or the like."[18] Soon then Whewell's axioms of morality—such as right and wrong, benevolence and justice—would be taken for granted and as equal to general axioms in geometry and mechanics. Nonetheless, he removed the analogy between geometry and morality in the second edition of his *Elements of Morality* in 1848.

Morality at Cambridge was being destroyed by the system of coaching. Hare reported the need to reduce "those two terrible evils in our system, the practice of private tutors, & the use of emulation as the one great spur to the acquirement of knowledge." The strength of Oxford was the fact that the university had a "vision," while Cambridge had simply "sunk." He warned, "We shall continue to sink, unless we get rid of our system of drilling & of our morbid stimulants, & adapt a system which will call forth a living power and train our students to walk without leading strings." The man to do this, he advised, was Whewell, greatly aided by the able Worsley.[19] Whewell, on being appointed vice-chancellor toward the end of 1842, had tried to reduce the influence of private tuition and to increase the role of lectures. However, this was proving very hard and, as was typical for Whewell, when things were not going his way he became irritable. The behavior of students in the galleries of the Senate House during ceremonies was troubling him. He issued instructions to college tutors to ensure good behavior from students. He also wanted pigeon shooting on the outskirts of town banned, along with rowdy student dinners at taverns.[20] A system of morality could not be grounded without true belief in God; this was a transcendental quest implicit within obtaining knowledge and defining universal legal laws. Moral activities were intrinsic to following divine legislation.

In May 1847 Whewell, now at one with his late friend Rose, preached to a captive audience of the Philanthropic Society at St. George's Field, Southwark, on the Christian's duty toward transgressors: Jesus taught his disciples that "The kingdom of God is within you." The gospel revealed that all men were alike and helping people to find faith was a much higher calling than simply trying to relieve the material distresses of this world. "The disciples were of one heart and mind; and this oneness of heart and mind at once swept away the boundaries of division of property and distinction of condition."[21] The instructions given by the apostles revealed the divine foundations of human society, while "the powers of the human intellect" were given to us by the Creator to "ennoble and purify man's nature"—it was here one found one's moral path.[22]

Finding salvation and preparing for eternal heaven were a regular theme of Whewell's sermons throughout the 1830s and until his death in 1866. This life, he told his flock of students at the Trinity Chapel in March 1843, was a short period and platform to the next world. "That we are hastening on to another world of good or evil; another form of God's government awaits us, near at hand." As such it was essential that "the eternal must govern the temporary." Indeed, it was sinful to think only of this world and be intoxicated by the human senses or abstract human rationality.[23] "Purification," he later emphasized, was crucial and could come only from within.[24] This was central to Whewell's education of Trinity students.

It was through imaginative pious men that ethics was revealed in each epoch; this morality was synchronous with those that also secured true knowledge of the world. England, as the world's cleverest and most civilized country, had to ensure it spearheaded the spreading of the true word; progress in true knowledge went together with the right morality.[25] This was why Whewell spent so much time establishing and defining moral philosophy at Cambridge—as far as he was concerned, the heart (especially Trinity) of England, intelligence, and global civilization.

Whewell first systematically made this argument in the publication of his four sermons, *On the Foundations of Morals* (1837). A primary target of the work was once again the moral utilitarianism of Paley and his sensory intelligent design argument. The treatise was also dedicated to Hare and harked back to their days together at Trinity: "In the main purpose of my sermons, I know that I shall have you for my favourer, for I had you for my forerunner."[26] In the first sermon Whewell outlined the role of Christian preachers, namely "to lead men to love and glorify their heavenly Master, and to live suitably to their Christian profession." He drove his message home by clearly drawing on the Gospel of Matthew and Romans 1:20, the key text for the possibility of a natural theology: "For the invisible things of Him from the creation of the world are clearly seen, being understood by the things that are made; even His eternal power and Godhead; so that they are without excuse."[27] On first reading, Whewell explained, this could support a Paley view that the divine Creation could be discovered via "the visible and material creation." However, Whewell had to deprive Paleyans of the support of St. Paul, and thus preached, since it "was not alone the power and skill of a divine artificer, which he held to be discoverable in the constitution of the created world, but the holiness of a divine lawgiver, the justice of a divine judge."[28]

The key was consequently to know where the "eternal Godhead"—the religious standard—could be found. This, of course, was within. "It cannot be in the mere world of matter, or even of irrational life. Stones and trees speak not to us of the righteousness of the Artificer." Animals and men are brilliantly designed but tell "nothing as to a *moral* author of the world." The answer was to be found in our "conscience"—where revelations in the Bible revealed what nature could not. In short, "there is a world within the heart of man, which the Creator of all things has made, as well as a world without." The most important works of God are "a fixed constitution of internal powers" (the soul). It is when we dwell on "this internal world of impelling and regulating principles, this mental constitution of man, this law written on his heart,—are we not led to a nobler view of the creation and of the Creator, than any aspect of the mere world can offer?"[29] It was, after all, what gave us life. God appears far more magisterial as "the ruler and lawgiver of the moral world, than when we contemplate the earth as his footstool and the sky as his canopy." Whewell closed his sermon with the following edict: "It is our duty—may it be our privilege!—it is our appointed cause—may it be our path of blessedness!—to find, within and without, in the book of revelation and of reason, in the original structure of our souls, as well as in the blessed provisions for their redemption and sanctification, evidence of Him."[30]

In his second sermon Whewell attacked all those who denied the existence of the soul and ridiculed the view "that the decisions of conscience may be resolved into the impressions produced in us by the experience and expectation of pain and pleasure." His target here was Benthamite utilitarianism and its penetration and destruction of so much of Britain's traditional constitution and customs.[31] Our senses were secondary to the internal laws of spirituality—"the very life of our souls, and without which even Christianity would have no materials for her task of the restoration of our nature" (after the Fall). This "moral power" needed "exercise and patient contemplation" to "form a system of solid doctrine." To understand "the law written on the heart, he must turn his eyes within, and use such care and patience as befits one trying to decipher the record of a concealed treasure."[32] The corruption of Adam did not mean we still did not have a glimmer of God's blueprint. Morality, and therefore Christianity, should be made central to any worthwhile Trinity and therefore Cambridge curriculum: "The practical student of Christian morality, constantly employed in choosing the good and rejecting the evil;—in directing himself by the voice of his conscience within, which becomes clearer and clearer as he listens to it;—can no longer doubt that he has a conscience within his breast, or that there is moral good and evil in human actions." By establishing such a Christian morality, the prevailing decay of society would be stemmed, and the correct ethics reinstated. Man's "inner being" would "be illuminated by a sunshine which gives at the same time light and warmth. If we could reach this condition, *how little should we need codes and systems, reasonings and doctrines*!" (my italics). The conscience of the Christian was and should always be guided by Scripture, the Anglican Sacraments, and accompanying moral duty.[33]

Whewell claimed he was induced to preach his four sermons to offer an alternative to Paley and erase "the evils which arise from the countenance thus afforded to the principles of Paley's system." These were "so great," he continued, "as to make it desirable for us to withdraw our connection from his doctrines without further delay." Paley's work was responsible for "a very large share in producing the confusion and vacillation of thought respecting the grounds of morals, which is so generally prevalent in England, even among persons of cultivated minds."[34] Hare too wanted everything to be done to help deliver "us from the *mad pursuit of material objects*, which enslaves—from the mad pursuit of factious objects, which renders asunder—the heart of the country" (my italics).[35]

For now, Whewell, as at Oxford University, advocated Paley's replacement with Butler and the joining of natural religion with revelation. Butler's *Analogy* (1736) taught the dangers of speculative reason and underlined that human reason alone was not the origin of knowledge.[36] Looking at man-made watches and other mechanical contrivances told us nothing of the holy spirit and men's hearts. However, the bishop's system of morals needed to be updated for nineteenth-century purposes. The person to do this would be Whewell who, on becoming professor of moral theology or casuistical divinity at Cambridge (1838–1855), undertook to replace Paley with his own works on the subject. He eventually achieved this by arguing that man's moral "fundamental ideas" were as real as physical "fundamental ideas" such as "time" and "space." Whewell shifted the utilitarian morality that the individual's calculation of consequences was the source of right and wrong, with the argument that man's true consciousness was the voice of God; human conscience, indeed life, was nothing without true divine faith.[37]

This belief lay at the center of the TC—it provided, as Hare put it, "the best prospect for the illumination of all our faculties, as well as the groundwork of a true human morality." Hare wanted "to lay the foundations of morality deeper than those who are merely aiming at right acts can lay them."[38] Forging the right human morality was vital: "There is no stability for morality, except in Faith." Humans, at least the English Anglican elite, needed to find pure faith and spread the word among the unchosen to ensure social stability—be it the urban middle class, working class, or savages abroad: "May such Faith be granted us! May we ever acknowledge with our minds, and feel in our hearts, that God is the only eternal reality, and that all things else are only real, so far as they are in Him!" It was crucial to let faith lift the heart, will, and understanding from "those which the senses pamper and drug us." The prevailing emphasis on the visible world neglected the most important region of faith that was invisible and within. This had serious implications for morality: "all the primary principles and ideas of morality belong wholly to Faith," and "never come within the ken of the senses" or pure deductive human reason. All the TC shared this belief.[39] Whewell had earlier written to Jones on the spiritual part of man, "the minds that feel a conviction of principles of unity as yet undetected, that believe in the existence of truths wider than they can limit by phrases habitually current, and that assert to the possibility of a connexion among laws that seem far asunder, while they acknowledge their ignorance what the connexion is; these are minds which have the best chance of discovering new principles and new generalisations and such habits of thought lead naturally to the persuasion of a supreme principle of unity and connexion."[40] Reliance on "mere human knowledge" eventually was of little use and, moreover, was dangerous in diverting us from the true divine path—commencing with inner purification—toward the next eternal world.[41]

In 1849, after most of the European Continent had experienced political revolution, Whewell gave a sermon in the Trinity Chapel on the need, more than ever, to get as close to God as possible: "We must walk daily with him in our inward soul—seek his help to purify and elevate, to warm and enlighten us . . . that we may be advancing towards the gate of heaven, even before death knocks at the door."[42] Thirlwall also gave a sermon that year at Llandingad Church in the beautiful Carmarthenshire countryside to celebrate the opening of the Welsh Educational Institute. Such institutes should be "a training-ground for exercising and unfolding the noblest energies of our spiritual nature." Now sounding more like Rose, he emphasized the need to direct students to "loftier aims" that respected the Godhead and historical tradition rather than constantly seeking innovation. Clearly the events in Europe were a demonstration of the shocking repercussions resulting from neglecting our souls: "We live in eventful, changeful, critical times. We have just witnessed the revolution of a hurricane which has swept round the greater part of the civilized world, levelling ancient thrones and dynasties, uprooting time-honoured institutions, and shaking the foundations of civil society." Britain had so far escaped this destructive storm but now needed more than ever to be on guard. A key part of this would be to ensure the cultivation of one's soul within appropriate educational institutions—with the study of ancient languages vital to this process. Without this, "in every system of liberal education," the whole of the past would be illuminated from the present. This would eclipse "all sure guidance for the future, and to throw society back into a state of barbarism . . . to lock up the fountains of revealed truth, and to reduce the sacred oracles to a sealed book and a dead letter." The end aim of a liberal education should be to get as close to God and to love God—to gain salvation and prepare for the forthcoming eternal heavenly world.[43] This was an unmistakable reference to sacred Scripture and God's special revelations in the Bible.

Morality was central to Whewell's later explanation denying the plurality of worlds in 1853. Man's relationship to God and intelligence was unique in the universe. It was important to prepare man's soul for eternity and that meant synchronizing as closely as possible to the divine mind.[44] Morality was also what made possible the ability of man to form societies (unlike animals).[45] The ideal civil society therefore always included an ethical base structured by true Christian religion taught by a National Church. Throughout their time at Cambridge, the curriculum and process of teaching was always uppermost in the minds of the TC. This will form the subject of the next chapter.

CHAPTER NINE

SPACE, TIME, AND THE IDEA OF A CAMBRIDGE EDUCATION

> I think that the blessing of having a Church such as ours, so interwoven with the Constitution, so beneficial in its influence upon religion, morals, manners, and politics, is so great that we can never enough value it, and ought not to trifle with it. I think that this Church, having been so interwoven with the spirit of the country, must be continually identified with that spirit by the prevalent system of education, and that when this ceases to be done, the Church cannot but speedily fall—which would be the greatest evil the country could suffer. I think therefore that it must be the business of our statesmen to secure the permanency of a sympathy between Church and education.
>
> Whewell, March 10, 1840

Whewell was worried about the future of the Church and the support for its "interwoven" relationship with knowledge, morality, education, and architecture. He reflected in one of his personal notebooks, "How is it that we have undertaken, after the lapse of six thousand years [God's creation of humans and the world], to manufacture a morality of our own; to decide by a *cold calculation of interest, by a ledger book of profit and loss*, the preference of truth to falsehood, of piety to blasphemy, and of humanity and justice to treachery and blood?" (my italics). The driving utilitarian-commercial imperative of basing everything on human reason, good bookkeeping, and counting was driving Whewell mad.[1]

Charles Dicken's Ebenezer Scrooge and Thomas Gradgrind loomed large for Whewell too. It was his divine task while on earth to preserve Trinity and a Cambridge education from such a disastrous intellectual, irreligious, materialistic, and morally empty invasion. His fear of such a prospect was once again making him extremely petulant. "This year," Hare warned him, "you seemed to me to have become quite as vehement as ever." His friends knew he had a temper and rage that could do more harm than good, and he needed to reign it in.[2] Space and time were two leading fundamental ideas and illuminated by sacred geometry. They were an intrinsic component to the idea of a Cambridge education.

George Macaulay Trevelyan, a later master of Trinity, succinctly captured Whewell's approach to the curriculum via his clever changes. On the surface he "made sustained and successful efforts to broaden the scope of academic study." In practice, however, Whewell's central objective was to defend and prevent innovation at Trinity and any meddling of university statutes.[3] Whewell restricted the extension of scientific teaching, particularly the new sciences such as geology, into the university curriculum but expanded the teaching remit to accommodate a deepening of moral and Christian teaching. Like Arnold at Rugby, Whewell sought to emphasize moral and intellectual guidance into the curriculum at Trinity. This meant resisting whenever possible the march of utilitarian rationality.[4] The self-made Whewell was a fervent elitist and traditional Tory defender of the status quo.

Already by this decade Whewell was vigorously promoting a new curriculum that would eliminate from the university precisely that destabilizing and ungodly pursuit of innovation that people like his former colleague, the Cambridge Lucasian Professor of Mathematics Babbage (1828–1839) epitomized. In Whewell's *Thoughts on the Study of Mathematics, as a Part of a Liberal Education* (1835), he recommended geometrical reasoning because "such a kind of study will require and occasion more thinking than the mere application of technical rules." Further, he scoffed, a "scheme of study which escapes or tries to escape the labour of thinking, will answer none of the purposes at which we ought to aim." Whewell concluded, as we saw in an earlier chapter, by offering the extreme solution of abolishing algebra from the Mathematical Tripos: "Let the knowledge of Algebra be required no longer, for I can hardly believe that this part of our mathematical teaching is of much value in any point of view." To discipline the mind through such a system was to force it into an iron cage, requiring it to interpret the world in a way that was insensible to moral evidence and to "those finer perceptions of fitness and beauty, in which propriety of action and delicacy of taste must have their origin."[5] It was to excise the divine essence of life.

Hare had also in his early Commonplace Book pondered the place of algebraic mathematics in education, which he described as "heartless" and functioning via a callous form of reason in which the mind "wanders . . . through a dark and subterraneous passage." By contrast in poetry "every step is clear, conviction is immediate and complete."[6] This could have been straight out of Coleridge, who saw this turn to mechanized understanding as the great sickness of the century. He argued it was a direct product of the growing dominance of industry and utilitarianism. The Antichrist had hijacked understanding, and the higher domain of reason needed to be reasserted, that is, the divine and eternal basis of truth.[7]

Babbage was never part of the TC and clearly opposed most of what they stood for. No wonder he fled Trinity for the more accommodating Peterhouse, following sharply to London. Despite returning to Cambridge as Lucasian Professor he never took the position seriously, at least not how Whewell and others demanded, and left—not that he had ever lived permanently in Cambridge—once again for London eleven years later. He could never and would never fit the mold of a Cambridge man—let alone a Trinity man. His work on the Difference Engine during the 1820s was an overt attempt to build his materialistic philosophy of the mind into a machine, and he referred the process of the engine's carriage mechanism as similar "to the operation of the [human] faculty memory." By the early 1830s he believed he could now build the human faculty of foresight into an Analytical Engine: "At last having exhausted, during these years of labour, the principle of successive carriages, it occurred to me that it might be possible to teach mechanism to accomplish another mental process, namely—to foresee . . . as soon as that was attained, the next step was to teach the mechanism which could foresee to act upon that foresight."[8] Calculation and perhaps all human reason could one day be mechanized. All this had massive implications for notions of intelligence and therefore education.

At the start of the nineteenth century almost all in social authority agreed Christianity should be central to an education, but the incompatibility between the accepted principles and doctrine of Christianity made such a quest deeply problematic and political. This led by the end of the decade to an actual questioning of whether religion should even be in education at all. One of the leading educational reformers was Henry Brougham who by this point wanted to avoid conflict by cutting religion out altogether and therefore eluding the need to define a denominational identity. This Whig thrust for educational reform was always a prominent concern for the TC.

"FANCY MEN & PRIVATE TUTORS"

This unwelcome rise of a utilitarian approach to educational reform was right on the doorstep of the TC. Not surprisingly they were greatly disturbed by the perceived mechanical and factory mentality being introduced via the coaching system at Cambridge.[9] By the time the TC had arrived at Cambridge, mathematics dominated teaching at all levels, and this was producing, despite legal issues, coaches or private tutors (typically college fellows but also many who could not be fellows because they were married) who charged a fee. The key to wrangler success was to inculcate a high degree of technical skills within a short period. This, argues Andrew Warwick, triggered a "pedagogical revolution" in undergraduate studies. Whewell became increasingly despondent with the role and aim of mathematical teaching, believing geometry alone should be the tool that brought students closer to their soul and an eternal future in heaven. It should not be a device, as algebraic analysis was becoming, to aid the industrial production of wranglers and subsequent fodder for advancing related research. This contrasted with a onetime Suffolk farmer and now leading tutor of mathematics, the "senior-wrangler-maker" and Peterhouse graduate William Hopkins, who fervently believed in the industrial coaching system.[10] The fact that Hopkins stemmed from Babbage's former adopted college, Peterhouse, would have made him even more disliked. An education should not be like a factory churning out men like Babbage but, rather, godly men like Whewell.

Whewell later pleaded in elitist black-and-white terms: scientific discovery was achieved "not by a mechanical method which levels all intellect, but by special efforts of the brightest intellects to catch hold of the [fundamental] ideas by which these laws of phenomena might be interpreted and expressed in more general terms." The whole history of Trinity's illuminati proved this—from Bacon and Newton through to Airy.[11] By contrast, for example, the Peterhouse farmhand mathematician Hopkins's work on glaciers was clear evidence of forcing algebraic analytical mathematics into a realm where it was not needed and made understanding abstruse.[12] Mere mathematicians like Hopkins should know their place and stay out of the realm of discovery. Whewell's close Scottish friend, James David Forbes, also a savant of glaciers, asked Whewell, "When will you be quit of the tyranny of the fancy men & private tutors."[13] Ironically for Forbes, it was precisely Hopkins's approach to mathematics and "drilling" that would nurture his onetime student, William Thomson, into becoming one of the most successful future industrial savants. Thomson was born in Belfast before studying at the University of Glasgow and then Peterhouse. Although the future Lord Kelvin despised the kind of metaphysics that Whewell expounded, he did admire his *Mechanics of Engineering* (1841). Kelvin shared Whewell's view that mathematics must be clear and visualizable, but completely disagreed with him that geometry met this criterion.[14]

Whewell assured Hare he was doing all he could "of the object you speak of—diminishing the influence of private tutors."[15] In the curriculum reform of 1848 Whewell clashed with several of his university colleagues, especially Hopkins, over the role of private tutors. This was all aired in the subsequent Royal Commission's investigations of teaching and curriculum teaching at Cambridge during the following decade. This would lead to tension between the Trinity master and the commissioners, who included Herschel, Peacock, and Sedgwick. Whewell firmly told them all to back off, that private coaching "enfeebles the mind and depraves the habits of study."[16] For all of Sedgwick's ranting, he clearly did not dislike the coaching system in the way Whewell and Hare did. Whewell feared that the excessive emphasis on competition via written exams, at the expense of the old oral disputations, was greatly undermining the student's intellectual and moral faculties.[17]

Whewell was not confident and acknowledged it was an uphill struggle; he earlier confessed to Hare that he had seen enough to suggest "the old universities cannot expect a continuance of the protection which they have been accustomed to receive at the hands of government."[18] He had done his best in 1845 to protect the system when he published his defensive treatise *Of a Liberal Education in General, and with Particular Reference to the Leading Studies of the University of Cambridge*.[19] At the time he told Jones that he wrote the book because "people were going on in a very thoughtless manner with regard to changes in our examinations which I hope I shall stop."[20] Whewell received support from Astronomer Royal Airy, who agreed that the emphasis at Cambridge should be on "the great standard works of all times rather than to the last steps made." Crucially, they should be on guard against the Babbages of this world: "On the principles of d-ism, in opposition to the dot-age of the University."[21] This was a play on Babbage's original justification for forming the Analytical Society back in 1813. Airy did not like Babbage or, at least publicly, his mechanical philosophy. Whewell's and Airy's mutual hatred of Babbage was clearly a powerful bond between the two Trinity men.

Already in the latter half of the 1830s Airy had worked behind the scenes advising his former Trinity friend, Sheepshanks, in defending the instrument maker Edward Troughton against a legal challenge by the London astronomer and former surgeon James South. South was good friends with Babbage and Herschel—all had been part of a London-based group that formed the Astronomical Society back in 1819. South had employed Troughton to construct an equatorial mount for his telescope. However, he subsequently declared it faulty and wanted his money back. At the trial he called Babbage as an expert witness. Sheepshanks was furious and attacked the fallen Trinity man's credibility, helped no doubt by Airy behind the scenes.[22] The judge at the trial was another former Trinity graduate and onetime member of the

Analytical Society, William Henry Maule, who came down in favor of Troughton. In 1835 Maule became counsel to the Bank of England after succeeding another Trinity man, James Scarlett.

In 1842 Airy upped his dislike of Babbage and advised Peel's government not to continue funding his calculating machines, declaring them useless.[23] By contrast, Airy fully supported Whewell's argument on "the tenor of your remarks on the mind-destroying effect of analytical process." The treatment of numbers by symbols was, however, a very powerful tool. Its use should remain but, Airy legislated, confined to "a few persons among whom I class myself who have very severely disciplined themselves in the examination of the evidence of every individual step." In general, however, "as regards the discipline of the mind it is greatly inferior." It made the mind nothing but a machine with each symbol merely acting like a cog.[24]

By this period the old established texts and certainties of Euclid's *Elements* and Newton's *Principia* were being usurped by new competitive examinations in advanced analysis. Whewell was deeply concerned over the impact this mechanical and industrial approach to teaching was having on the students' faculties. A university was there to produce the country's ruling elite and not a collection of specialized professionals; mathematics was there to demonstrate known principles in, say, gravitational astronomy and optics and not new speculative doctrines missing a binding fundamental idea. Thus Whewell worked relentlessly to remove or prevent subjects such as geology, electricity, magnetism, and heat entering the Cambridge curriculum.[25]

In November 1848 Whewell's great friend, Forbes, also waded in with a lecture at the University of Edinburgh defending a liberal education. He slammed the emphasis on exams and simply cramming the memory rather than cultivating the understanding; this was simply superficial learning that would end in atheism. Students had to exercise all of their God-given faculties. One could obtain the conclusions, say, of mathematics and natural philosophy but remain entirely ignorant of following the demonstration that defined the road to the conclusion. Forbes generously cited Whewell to support his argument and concluded, "When study is divorced in any measure whatever from reflection, it degenerates into what is vulgarly, but expressively, called *cramming*; an effort of memory. Or at best a kind of mechanical expertness, which simulates real knowledge, often so effectually as to deceive not only bystanders, but the learner himself."[26]

Despite his later support of the coaching system, perhaps more than any other recent publication about a university education, it was actually Sedgwick's earlier brilliant polemic, *Discourse on the Studies of a Univer-*

sity Education (1832), which had inspired Whewell and the Trinity tutors on the need to safeguard its traditional pedagogical role. Indeed, the *Discourse* was published due to an official invitation by Whewell just after Sedgwick—a popular figure among undergraduates—had given the work as the university's Commemoration Day sermon. Whewell told Hare that the lesson was "as I think you know, a beautiful and profound dissertation as he delivered it; and having been rethought and rewritten since it is much more beautiful and coherent; and along with certain notes which he has appended to it, it forms an essay upon philosophy, morals, and academical education which it will delight you to read."[27]

Whewell proudly announced to Rose that at last Cambridge had a knight to defend the true faith in this year of profound change (the Great Reform Act)—such a person was needed more than ever. Sedgwick "has declared war against both Paley & Locke. This puts them in a different footing in Cambridge from that on which they have hitherto been; for though opinions to the same effect were in very general circulation in the place, they were never I think clothed with anything like an authoritative expression before." It was now the time to introduce a new system of ethics founded on true morals. He encouraged Rose to apply to Durham University for the vacant chair in divinity where he too could begin work on such a new ethical system.[28]

In the preface to his *Discourse* Sedgwick condemned the "utilitarian theory" of morals not just for "false reasoning," but because of the "degrading effect" it had on "the temper and conduct of those who adopt it." Ethics was reduced to mere calculation. What was needed, he declared, was "the establishment of a system of moral philosophy to banish it." As we have seen, this was taken up later with gusto by Whewell. A Cambridge education should, according to Sedgwick, create a "well-disciplined mind" that learned "to see everything with an eye of faith, so as to find, in all the dispensations of Providence, a motive for the exercise of holiness." It was within Cambridge's, especially Trinity's, robust Anglican walls that the world's greatest philosophers had nurtured their souls. As well as Bacon and Newton, this included men like John Ray and Isaac Barrow.[29] All knowledge, rightly so-called, is moral. The laws of morality were as firmly implanted as physical laws; a divine relationship between humans and the physical world.[30]

For Sedgwick, like the TC in general, it was vital that a Cambridge education cultivated one's soul: "In some of our capacities, we may perhaps exhibit a faint shadow of a portion of our Maker's image; but in the reasoning power, of which we sometimes vainly boast, we bear to him, I believe, no resemblance whatsoever." It was through the study of "Newtonian philosophy" that our "moral powers and capacities" were developed and taught "us to see the finger of God in all things animate and inanimate." This ultimately prepared the mind to be obedient "to the divine will." The external world revealed God—to and through a few chosen pious Christian men—"by addressing the imagination, and by informing the reason. It speaks to our imaginative and poetic feelings, and they are as much a part of ourselves as our limbs and our organs of sense."[31] Whewell was adamant and told Hare, "If we do not" conform our whole character to the divine rule, "we offend; and any special acts of sin only betray this want of perfect discipline in our character."[32]

Sedgwick condemned those who tried to erase the truth of their Creation, namely that man came into "being within a few thousand years of the days in which we live." The idea that man had evolved from the transmutation of species was "a theory no better than a phrensied [frenzied] dream." The first cause contained "a provident and active intelligence." God had not simply created "the world and left it to itself"—albeit programmed—as deists like Babbage and numerous others, including Herschel, believed. As with most of the TC, Sedgwick underlined the importance of studying languages and all the "great works of imagination which have become a pattern for the literature of every civilised tongue." It was vital "to follow the track of those who have adorned the history of mankind—to feel as they have felt—to think as they have thought—and to draw from the living foundation of their genius." The bedrock of the present innate in ancient language and scrutiny would reveal realities that were apparent before becoming words—the latter being "signs of thought." Those, such as Paley, who claimed "ancient systems" obscured "moral principles" were completely wrong; indeed, the truth was the contrary. History in general was fundamental to knowledge of man and was the equivalent to "physical experiments" and, as such, the gateway to "our knowledge of the laws of nature." An Anglican religious nature was imperative to both objectives.[33]

It was only through being in touch with one's soul that innate divine ideas were unlocked and released to guide humans. This was completely ignored by wrong-headed thinkers such as Locke, Hume, Paley, Bentham, Mill, and Babbage: "What would be the value of the senses were there no sentient principle within? And where would be the use of teaching were there no inborn capacities to the soul to apprehend and to be acted on?" It was our imaginations that elevated man above beasts. Lockean and utilitarian morality was, for Sedgwick and the TC, "an outrage on" man's "moral nature." A soul was vital, since "without an inherent moral capacity, without a moral sense placed in the breast of man, by the same hand that made him, the science of moral philosophy was not, I think, the shadow of any foundation whereon to rest."[34]

Everything corrupt in Paley, claimed Sedgwick, derived from Locke's *Essay on Human Understanding* (1689). For Locke, he sighed, "God, as far as regards the interest of man, must be benevolent, malignant, or indifferent. This is the fundamental proposition of his moral system." For Paley too our morality depended on our "following a circle of life, measured by the standard of utility." Sedgwick was scathing, and one can imagine the satisfaction the TC and others felt when he vomited a venomous rant against utilitarians and the prevailing materialist spirit: "*Virtue becomes a question of calculation—a matter of profit or loss*; and if man gain heaven at all on such a system, it must be by arithmetical details—*the computation of his daily work—the balance of his moral ledger*" (my italics). This was outrageous to true belief: "Utilitarian philosophy and Christian ethics have in their principles and motives no common bond of union, and ought never to have been linked together in one system." By doing so Paley had proved more destructive to the true faith than men like Bentham. This philosophy destroyed "the dominion of the moral feelings" and offended both "the law of honour and the law of God." Possibly even worse than the sensory bigots were the abstract analysts. Where God was expunged it led, as the relatively recent events of France showed, to a "school of fanaticism." Indeed, Sedgwick sneered, "Had they [the French] accepted, even in the humblest degree, the doctrines of the religion of Christ, they never could have made such portentous errors in estimating the moral character of man."[35] This was why Sedgwick was so important to the TC's educational concerns.

For Sedgwick, Whewell, and the TC in general, science had to be assimilated into the traditional structure of a liberal education, which should be in equilibrium between new and old knowledge. As Whewell implied in 1840, "In framing scientific terms, the appropriation of old words is preferable to the invention of new ones."[36] Whewell's moral philosophy was designed to restore the balance of a Christian liberal education through the correct training of the Cambridge, primarily Trinity, mind. He used it both to legitimate a particular stance (as opposed to the sole use of sensory observation or sole use of pure human reason) and to show that science did not progress through sudden changes with the past, but rather accommodated old claims within a new perspective. In other words, it was organic—in the Coleridgian sense—and not built on unprecedented abstract axioms.[37] Sensory and mechanistic teaching at Cambridge that drew on Locke and Paley should therefore be surgically removed, and—especially French Comtean positivism—aggressively protected against.[38] As Jon Agar shows, mechanism was equated to simply the unthinking following of rules. This, as we have seen, was argued as akin to Continental algebraic analysis. The shift at Cambridge to a mechanistic system of coaching mathematics was part of a general movement away from the cultivation of a gentlemanly elite.[39] Whewell dismissed Comte as having "opinions on the philosophy and history of science" that were "of no value." He represented the "character generated in France by the prominence given to the study of mathematics in the last generation," especially Lagrange, Laplace, and Lacroix.[40]

By the end of 1847 Whewell tried to persuade Hare to be a candidate for the vacant chair in theology at Cambridge but failed. By this point Hare's Anglican income at Hurstmonceux was too important to his material quality of life.[41] Instead, the Regius Chair of Theology went to the former Trinity graduate and professor of classical and general literature at the East India Company College in Haileybury, James Amiraux Jérémie, in 1850.

A huge victory was achieved in 1855 when a university regulation quietly announced that an acquaintance with Paley was no longer needed.[42] Sedgwick, Whewell, the demoted Thirlwall, the recently departed Hare, and the deceased Rose would all have felt very satisfied.

GOTHIC ARCHITECTURE AND THE RESTORATION OF TRINITY COLLEGE

The space of a Cambridge education was both mental and physical. Whewell was a master of preempting future reformist attacks on his beloved college and university. His version of Anglican belief informed his approach to education, institutions, philosophy, and everything he pursued—although not perhaps his private financial investments. Like his predecessor, Christopher Wordsworth, he was also instrumental in redesigning the college architecture. In both cases, they drew on a Tudor-Gothic vein. A sacred place of learning like Trinity should spatially reflect its organic teachings. William Wordsworth described his own literary work as a Gothic cathedral, and reminisced that his study of geometry at Cambridge had been the most valuable part of his education. As early as 1818 Whewell believed that the most important aspects of our manners and constitutions were derived from a Gothic past; the correct morality created the right architecture.[43] Under Christopher Wordsworth's reign, New Court (1823–1825) was built in the Tudor-Gothic design by the Gonville and Caius graduate William Wilkins, who also went on to design the East India Company's college at Hailsbury and several other Cambridge college buildings.

Gothic architecture had grown organically, with each of its parts linked together and united. For Whewell, this form of divine architecture had first commenced in the German states, with the adoption of the pointed arch of equal height and different widths to satisfy the requirements of vaulting.[44] Moreover it was designed by a divine language, namely Euclidian geometry, with its array of circles, straight lines, and equilateral arches, that is, the pointed arch. The whole building was symbolic of God. In Britain the architect George Saunders had identified the role of pointed arches in giving greater flexibility for spanning different widths in the construction of ribbed vaults. However, it was the architect Thomas Rickman who informed most scholarship on medieval buildings with his publication in 1817, *An Attempt to Discriminate the Styles of English Architecture, from the Conquest to the Reformation*. Rickman scrutinized individual building parts and located them within four distinct successive styles: Norman, Early English, Decorated, and Perpendicular. This taxonomic approach clearly informed Whewell's and his Cambridge colleague Robert Willis's interpretation of architecture. From 1837 the latter—a Gonville and Caius graduate—was appointed Jacksonian Professor of Natural Philosophy, which he retained until his death.[45]

Whewell wrote, "Amongst other objects I was naturally led to search for evidence which would throw light upon the origin of the pointed arch."[46] This type of arch for Whewell was the central feature that kept the whole structure united. By contrast, for Willis it was the arch itself and not the pointed aspect that was central and did not represent the divine symbolic aspect as it did for Whewell.[47] The pointed arch allowed less enormous walls and provided sizable openings for windows that could be placed much nearer together. This allowed a far brighter spatial interior—all enhanced by significantly taller edifices supported by flying buttresses. Consequently, worshippers were, literally, much closer to the light of God. The whole verticality of the building pointed to an eternal future in heaven, just as induction and fundamental ideas, rather than deduction or just sensory experience, revealed divine truths.

The approach undertaken to the history of Gothic architecture was induction, and for Whewell the key was to locate the fundamental idea that linked the observations together. In this case the idea was verticality with the pointed arch at its heart. This verticality helped create a unified system but, also acted as an indicator toward heaven. The use of the pointed arch in vaulted roofs was a product, he argued, for the need to have arches of equal heights but different widths. The most developed and ideal examples, he claimed, were found in Germany. Those medieval churches were built for God and helped extend our understanding of Christianity.[48]

Whewell described Rickman as "a little, round, fat man, with short, thick legs, and a large head," who walked very slowly and moved randomly as something continually caught his eye. Nonetheless he liked him "very much. He is very good-humoured, and very intelligent and active, and I see more by travelling with him than I should do alone, besides understanding the architecture much better." In twenty-five days they examined 130 churches. Whewell did another tour with Rickman the following year around Northamptonshire to examine churches his friend thought older than the Norman Conquest.[49] For Whewell, Rickman had set the standard for establishing architectural nomenclature. Medieval structures in this sense were no different from the philosophical approach to classification in the natural world. Whewell likened it to, for instance, botany.[50]

Faultless ideal Gothic architecture represented the perfect idea of verticality. It played a part in the structure of the whole building so that as the architecture developed so too did the spread of verticality, from the buttresses and the windows to the building itself. Indeed, the flying buttress was for Whewell the essential aspect in completing the ideal Gothic building—it made possible the full-extent of applying pointed arches. He described the Gothic arch as a "silent finger" that "points to heaven." The origins for Whewell of the "complete Gothic" was Germany, with Cologne Cathedral its perfect embodiment. He wrote to Jones in 1829, "Cologne better than anywhere else the introduction of the pointed arch as an artifice of vaulting, and afterwards into the details, might be abundantly exemplified," and concluded that Cologne Cathedral was "the most . . . completely formed Gothic."[51] Two German architectural historians, Georg Moller and Sulpiz Boisseree, had already claimed that the origin of Gothic architecture was Germany and illustrated their work with drawings of the facade of Cologne Cathedral. Whewell was certainly aware of their work. Indeed, he met Moller at Darmstadt in 1829 and referred to Boisseree as his "old architectural friend" in a letter to his sister in 1849.[52] For Whewell this cathedral was also built during the height of complete Gothic architecture in the thirteenth century.[53] He told his sister in 1849 that he was "inclined to think [the cathedral,] the finest building in the world (for it is the finest Gothic cathedral and a Gothic cathedral is the finest kind of building)." Although by 1857 he now thought France was probably the leading country "in *this* history of art," that is, architecture of the middle ages.[54]

In 1842 Whewell, under the guidance of the architect Anthony Salvin, set about restoring some of Trinity's architecture. Together they discovered the remains of where the oriel once existed as part of the front Lodge. At the same time Whewell's friend, the ex-Trinity High Churchman and Cambridge Camden Society leader Beresford Hope, donated £300 to the restoration. The money was "to be devoted to the purpose of restoring the lodge the oriel and mullioned windows, exactly as they stood before Bentley's alteration and thus in some measure giving back its antique character to the old court."[55] First, the sash windows from the east face of the Lodge were removed and replaced by Elizabethan oriel windows to ensure continuity with the Great Court. Whewell then had rebuilt and greatly expanded the two windows on either side of the Lodge to increase the light into the Elizabethan drawing room. He also had the Tudor oriel window restored on the east side of the Great Court in a polygonal shape, which had been removed in the late eighteenth century. The window on the west side was also removed and replaced by an oriel window that greatly improved the view of the dining room and drawing room.

Another undertaking on Whewell's list was the restoration of Trinity's Great Gate and the completion of its hitherto unfinished vault. The architect here was Willis.[56] The past should be accurately preserved in the material space of a place of Christian worship or an Anglican place of learning. In this case it was the outward spiritual and doctrinal embodiment of the college. The funding for the oriel window restoration had partly come from Whewell (£250) and Beresford Hope. The bulk of the capital, however, came from the college (£3,765); so when Whewell erected a plaque celebrating the restoration as his work, along with Hope, many Trinity seniors felt he had gone too far with personal praise. The inscription was quietly removed when the stonework of the oriel was repaired in 1965. Nonetheless, Whewell's legacy, just like his predecessor Wordsworth, is preserved in brick. The latter had been instrumental in expanding the college via the building of New Court in 1823. Likewise, through two lucrative marriages, Whewell had become a wealthy man and donated £100,000 to building two new Gothic-inspired courts, codesigned with Salvin, opposite the Great Gate that are known by his name.[57]

Whewell just could not stop meddling in university affairs, be it architecture, curriculum, university rules, and even the institution's art collections. For example, he upset the Fitzwilliam Museum Syndicate when, under his own volition as vice-chancellor, he altered the arrangement of the paintings on exhibition. He insisted they should be displayed by grouping them into schools, resulting in the resignation of seven of the eight Syndics.[58] The real impetus behind Whewell's meddling was his disapproval "of nude figures in a public gallery" and the potential damage this could do to young visitors. Consequently he had them removed.[59]

Whewell also became instrumental in the building of the chapel at Mill Road Cemetery. The Gothic-style Lodge, built in 1848, was perceived as too small for large funeral parties. Thus there was a clamor to build a chapel. Money was raised, and the person appointed to design the building was a well-known Gothic-revival architect who had recently done some work on Trinity, George Gilbert Scott, probably at Whewell's recommendation. Scott had earlier worked on the design of Gothic-style workhouses following the Poor Law Amendment Act of 1834 and had been appointed by Peacock to restore the architecture of Ely Cathedral in 1847. He submitted his plan for the chapel in 1851 but it proved too expensive. The committee in charge of building the new chapel subsequently took advice from Willis and Whewell. To save money Scott reluctantly agreed that the tall spire had to be omitted and replaced by a lead turret. Whewell had never been happy with this compromise and wanted it to resemble an Anglican parish church—an essential feature insisted on by the Cambridge Camden Society in their advice for church building. Thus, following the death of his wife Cordelia Whewell in 1855, he provided £250 for a tower and spire to be erected.[60] Indeed, he wanted her buried in the shadow of the spire as a monument to her life. He also planned, at this juncture, to be buried next to her.[61]

THE ANGLICAN CURRICULUM AND UNIVERSITY REFORM

For Babbage and other contemporaries the TC approach was regressive, irrational, and mystical, a conservative group committed to preserving the establishment and their privileges. The TC and their ilk had created a philosophy designed simply to ensure the prevailing status quo. The Drummond Professor of Political Economy at Oxford University, Whately, with one eye very much set on disrupting such a view, lectured to his students in 1831, "I believe it to be a prevailing fault of the present day, not indeed to seek too much for knowledge, but to trust to accumulation of facts as a *substitute* for accuracy in the logical processes."[62] This was a brave new era in which the human mind was teetering on emancipation from the shackles of the established Church and state. Not surprisingly, Whately was the choice of the Whigs to succeed William Magee as archbishop of Dublin in 1831 with a remit to try to reform the Irish temperament.[63]

For reform-minded men like Whately, it was all about human logic and the way inferences were combined and reasoned, as opposed to the accumulation of data and vague notions of divine fundamental ideas. Much to the TC's dismay, this "deductive savage" was thrust to the center of designing the government's church policy.[64] His appointment made him even more dangerous than others such as Ricardo, Bentham, Mill, and Babbage. Discovery and organization, in this reading, were nothing other than a mechanical, human-centric, and godless exercise—

although, Whately, of course, would not have seen it this way. He saw the inductive approach of men like Jones and Whewell as promoting a mist of disjointed observations and confusion. The basis of political economy was clear logical axioms, operating like an algebraic machine rather than a collection of vague inductive notes.[65]

It was precisely this approach that the TC despised. As Whewell privately reflected in his notebooks regarding the development of mathematics: modern algebraic analysis mirrored modern warfare, where victory was "obtained by working with masses of men" in which their role was "little more than the mechanical duty of fulfilling his [the master mathematician's] plan." In short, they played a part in the maneuvers to gain victory but did "not understand the battle." This contrasted with the ancient "Greek geometers," who acted like "their military countrymen" and overcame difficulties alone and in line "with the usual operations of human nature." In contrast, the modern analyst, "without pains approaching the difficulty, brings his formidable artillery of symbols to bear upon it, moves his alphabetical troops from position to position according to *the rules of his system* of tactics" (my italics).[66] Someone like Whately was the equivalent of the modern analyst in clergy clothing. He was a dangerous man. Whewell would also be thinking of Babbage's earlier work on *The Economy of Machinery and Manufactures* (1832), where the utilitarian wrote, "It has been shown, that the division of labour is no less applicable to mental productions than to those in which material bodies are concerned."[67]

The same sentiments as Whewell's were expressed fifteen years earlier by the Oriel tutor, Tractarian, and second son of William Wilberforce, Robert Wilberforce. All modern mathematics, he claimed, was the mechanical regurgitation of a few limited number of basic formulae. All that was required was "dexterity exactly like that of a banker's clerk in arithmetic or a sailor's in astronomy."[68] This was akin to the factory owner dividing the labor process so that each worker remained, as Marx claimed, alienated from the final material product. The world could not be reduced to mathematical reasoning operating to abstract definitions and rules; these arbitrary symbols blindly followed instructions that rejected fundamental ideas and the process of learning. Although not always obvious the landed wealthy, often had a great deal more in common with the skilled working classes in opposing the growing new capitalist mind-set.[69] There was a mutual elitism in both intellectual and material production.

Minds educated in a utilitarian and factory fashion, concluded Whewell, "have never been taught to grapple with real mechanical elements of pressure and reaction; they have been accustomed to trust the process of deduction to certain representative symbols and the rules of their combination." This might be a useful tool but not to those "who have no distinct conceptions of what they represent." Making, for example, "algebraic definitions alone the foundation of the Calculus" failed in any pedagogical purpose and simply led to unsubstantiated speculation. This was Newton's critique and reason for retaining, in general, a geometrical approach. As such these "analytical generalisations should not be fostered" within the curriculum at Cambridge. A system based on this form of study simply attempted to "escape the labour of thinking" rather than instill a proper education. This was why Whewell regularly concluded that algebraic analysis should be excised from the curriculum, since it was not of much value in mathematical teaching. Instead, mechanics and hydrostatics should be introduced into Cambridge's examinations for the bachelor of arts degree. The theory of gravitation discovered by Trinity's greatest natural philosopher, Newton, should lie at the center of at least their natural philosophical training. The ideal example of how this should be approached was demonstrated by the college's latest heir to Newton, namely the work of the astronomer royal and mathematician Airy. Whewell pointed to Airy's essay about gravity for the *Penny Cyclopaedia* and had already published it for Cambridge students (*Gravitation: An Elementary Explanation of the Principal Perturbations in the Solar System*, 1834).[70]

Whewell received enthusiastic support in his battle with the Cambridge commissioners over educational reform in 1850 from the Eton headmaster Hawtrey.[71] Indeed, backing was coming from all the places that mattered for social authority. Hawkins at Oriel intensified his desire to work with Whewell against the appalling reforming university commission: "Gladstone's lately published speech has much shaken me, & made me doubt whether it is not our duty to resist the Commission on public grounds."[72] Not surprisingly, when the vice-chancellor and master of Pembroke proposed a modification of the university's constitution in 1851, Whewell attempted to snuff it out before it had even gained oxygen.[73]

During this whole period Whewell was also becoming a significant landowner in his own right. By August Whewell found time to spend £2,100 from his marriage trust (£50,000) to Cordelia on a new house in Lowestoft and asked Jones for help purchasing the land nearby from the Board of Ordnance.[74] His private acquisitions did not, however, get in the way of his protection of Trinity and Cambridge. By early 1852 Whewell was back defending the university statutes from "the radical party trying to use the occasion so as to subvert the whole constitution of the University."[75] The following month Jones wrote, "I heartily wish they would make you dictator" over university reform. "The multitude of cooks will spoil the broth I fear & the unreasonableness of some of their crotchets is wonderful."[76] Later that year Jones was appointed to the Cathedral Commission that involved assigning duties to cathedral dignitaries and the task of diverting certain funds for other Church purposes.[77] This was all part of an attempt to provide a purpose for cathedrals. As secretary to the commission, Jones received a munificent salary of £800 per annum.[78]

Whewell's onetime contemporary at Trinity and recent author of a biography of the poet John Keats, Milnes, tried to reassure his old friend that he did not think the reforms sought any ill will toward the college: "My notion of University Reform is to make Oxford like Cambridge & all Cambridge like our College, which I trust you will consider sufficiently loyal."[79] Whewell was not convinced. Meanwhile back at Trinity, he had more immediate issues to deal with. He heard from the "Boro [Borough] Police Office" that the Watch Committee had a report of foot passengers having water poured on them as they left the college chapel. Whewell was already extremely bad tempered over the attempt to reform the university. This event made him even more livid, and he spearheaded the hunt for the culprit, a certain Lionel Dymoke, who occupied a room in Great Court from where the water apparently emanated.[80] This for Whewell was a prime example of the immorality running riot in the country and now even in Trinity. It could, of course, just have been naughtiness, and it is an open question how far Whewell could distinguish between this and subversion.

Between being appointed master and Dymoke pouring water on passers-by from his Trinity digs, the rational march of "steam intellect" had become further assertive. Major developments in geology and natural history were raising yet more doubts about the veracity of God's word revealed in Holy Scripture. Even Thirlwall's former favorite student, Sterling, was becoming increasingly disillusioned with religion.[81]

Sterling had been dead for four years, but his resignation from the clergy, his correspondence with Mill, and his lectures at the Quaker-established Royal Cornwall Polytechnic Society had greatly upset members of the Church of England. Moreover, Sterling had earlier founded a London dining society in 1838 specifically to encompass as wide a spectrum of intellectual elite views as possible. After his death, Hare along with Thomas Carlyle and Mill continued the society and renamed it The Sterling. This convinced many Church members that infidelity had now penetrated the very belly of the Church. They launched hostile attacks on the society via their organ *The Record*. Subsequently several respectable members resigned after they failed to change the society's name—including the former Trinity student and archdeacon of Salop, John Allen, Samuel Wilberforce, and the Harrow- and Baliol-educated Henry Edward Manning.[82]

Hare realized his former student was a controversial figure and was determined not to leave his reputation to the likes of someone like Carlyle—despite his also being an enemy of steam intellect. Such men would, he told Whewell, emphasize all that was "painful" to him, "while those parts of his life & character which to me are so precious, wd. have been thrown entirely into the shade." This is what compelled Hare to write his memoir of Sterling.[83] This, in turn, led to Carlyle similarly writing a biography of Sterling published later in 1851.[84]

Meanwhile, Bishop Thirlwall was worried by the ongoing Rebecca Riots that had been carrying on throughout south and central Wales during the 1840s. The smashing of toll gates and the hostility toward tithes and Church rates in general had reached disturbing levels among the peasantry and had serious implications for Church income. Not surprisingly, unlike Jones and Whewell, a shaken Thirlwall later voted to repeal the Corn Laws in 1846 and a year later supported the Ten-Hour Factory Bill.[85]

After a year of monumental European political upheaval in 1848, Whewell told his devout Christian Scottish friend Forbes that he worried the revolutions could throw the world "back into the condition in which it was when we were born [i.e. 1790s], and have been grinding on for so many years to no purpose."[86] The year 1848 was not only one of European revolutions but also great Chartist demonstrations. Jones was in Haileybury worrying about these sociopolitical reformers. The government had sworn in thousands of constables comprised of all ranks from dukes to humble tradesmen for a forthcoming political rally of some ten thousand Chartists gathering for a meeting on Kennington Common, London. The new railway network allowed many to come from around the country. In addition to the constables there would also be many "Blue Bottles [the police] & red coats [soldiers]." Jones just hoped there would be no shots fired, as that would simply fuel the volatile situation and trigger another Peterloo.[87]

The Magdalene graduate and rector of Eversley, Kingsley, was also gravely concerned about Chartism and sought to do something about it. Maurice subsequently put him in contact with the equally concerned lawyer John Malcolm Forbes Ludlow, who had only just returned from France and was strongly informed by that country's socialism. He had contacted Maurice on the crucial need to Christianize socialism; the result was the creation, spearheaded by these three men, of the Christian Socialist Movement. Hare was also keen on the movement's objective and the need to confront a selfish individualism and commercial spirit, as well as diluting the distrust among everyday folk toward the clergy. He advocated the need for an alternative outlet to the Chartist movement that led to a meeting of Hare, Ludlow, Kingsley, and Maurice with the Cambridge University Press publisher and printer John William Parker. The result was the creation of a new penny weekly journal for the working-class Christian man, *Politics for the People.* Although there is no evidence that Hare contributed to the journal, there is an indication that Thirlwall may have.[88] This was plainly a counterrevolutionary ploy, conservative and top-down.

Hare was very enthusiastic over the project and told Whewell, "It seems to me a noble enterprise & one of the things most indispensable for the preservation of the whole fabric of English Society . . . we must fight the battle of truth & order against falsehood & lawlessness." If not, he groaned, "the glory of England will pass away."[89] Hare's mood later changed in 1853, when Maurice was dismissed from his post at King's College, London. The reason he was sacked was for his leadership in the Christian Socialist Movement and his alleged unorthodoxy in his *Theological Essays* published earlier that year—in particular, his denial of the eternity of punishment. The materialist philosophy, which was so undermining of Christianity and the Established Church, took the view that man was completely mortal. As such he did not possess a soul that the Scriptures taught was his passport to an eternal life. Those who rejected the soul also dismissed the idea there was eternal punishment. It was this that made Maurice so subversive. Nonetheless, Hare was furious and planned to lead a protest against Maurice's sacking. He may have not shared his view but was worried that all the good work his pupil and friend had achieved for the Established Church would come to an end. This, he argued, would lead to people turning elsewhere. Whewell declined to sign Hare's letter of protest.[90]

A SPIRITUAL AND IMPERIAL EDUCATION

Hare, like Thirlwall, Whewell, and earlier Rose, viewed England as a divinely chosen nation to spread Christian civilization. This mission needed to be constantly inculcated. At the jubilee of Simeon's cofounded Church Missionary Society held in Brighton (1849), Hare preached, "For us [England], in a temporal sense, may it be said, to us, above all other nations, has God given the heathen to us for an inheritance and the uttermost parts of the earth for our possession."[91] It was precisely by gearing all English institutions to the service of the Lord that the country had been protected from the revolutions that had torn Europe apart the year before. This position was being undermined by the march of a godless utilitarian "steam intellect."

Education and a distinct Anglican curriculum was ultimately the binding objective of the TC. Privileged Church of England public schools—along with Oxford and Cambridge—were, as they still are, the nurturing ground for England's domestic and global political leaders.[92] Britain, the TC believed, had risen to world supremacy through divine providence and the spiritual cultivation of its people (well, those in the elite), not via a utilitarian education. Despite his view of Dissenters and promotion of a broader encompassing church, Thirlwall, like the TC in general, was still deeply conservative and was becoming more so. In 1853 he was asked to give a sermon at the reopening of the select Anglican St. John's School in Hare's old parish of Hurstpierpoint. The school was originally established by the Magdalen Tractarian Nathaniel Woodard at another location in this small sleepy parish in the heart of Sussex. It was then rebuilt in 1853 and designed to resemble the collegiate system at Oxford and Cambridge; the complex has connected inner and outer quads with the chapel and dining hall adjacent to each other. Many of the school's traditions were directly drawn on one of England's oldest (private) public schools, namely Winchester College (est. 1382).

Thirlwall preached to the chosen flock gathered at the reopening that England was divinely chosen to spread God's "purposes and dealings throughout the world—to be the pioneers of God's way through the desert regions of the earth; to be the carriers of His Truth, and of all the social blessing which spring from it, to the rudest and most degraded tribes of our race." England's role—via elite private fee-paying schools like Hurstpierpoint and especially institutions like Trinity—was "to teach the [world's] nations how to live." This depended on cultivating "the quality of its men" to take up their position as "the legitimate organ of the national mind and will." The whole globe depended on "the intellectual and moral condition of this class," and to do so they had to assimilate the wisdom of former generations. As such, education, "above all English education—must be thus rooted in the past," and concentrate on the application of religion "to the heart and the conscience."[93] As Thirlwall later put it, "The Spirit of Truth is essentially one with the Spirit of Holiness," which both illuminates the intellect and purifies the heart.[94] By now Bishop Thirlwall was at one with Whewell and all his deceased colleagues who comprised the TC.

Whewell had made this argument a few years earlier: "Has England not taught the nations how to live?" He answered his question in the affirmative. This was by "an admirable combination of active and original intellect with unequalled practical sagacity and force of character."

English culture and education, he concluded, "has constantly impelled the progress of thought and institutions in Europe, while, at the same time, she has held back from extravagances and atrocities to which the progressive impulse has urged more unbalanced nations."[95] Moreover, this had all been done with the Established Church at the heart of the country and, despite the emancipation of Catholics and Parliamentary reform in 1832, with the sacred Constitution still just about intact.

However, thirty years after Whewell uttered these words and a year after his death, Thirlwall grimly told the Trinity graduate Reverend William Henry Lyttelton that the role of divine education had been supplanted by science, which was "at variance with it; for it has introduced man to the knowledge of a vast system of secondary causes, previously unknown and unsuspected, which seem to separate him from God, and to exclude the idea of the immediate divine presence and agency, which was the condition of his early piety." This was one of the most important issues of the day, he concluded, and needed reconciliation. The status of the English being God's chosen race was at stake. Thirlwall was elected an honorary fellow at Trinity that year, and spent his final years fearing for the breakup of the Church of England and its severance from the state.[96]

CONCLUSION

THE LAST DAYS OF THE TRINITY CIRCLE

> The whole history of the world from the creation of Adam to the present day, is one connected and coherent plan in the mind of God.
>
> WHEWELL, MAY 19, 1853

> There are some who rank Euclid as second in sacredness to the Bible alone, and as one of the advanced outposts of the British Constitution.
>
> JAMES J. SYLVESTER, AUGUST 1869

The core perspectives that lay at the heart of the TC reached their greatest unity during the 1820s and prior to the clash Rose and Whewell had with Thirlwall over the admittance of Dissenters in 1834. This was recognized by the Etonian, Oriel theologian, and Tractarian, Richard Hurrell Froude. In April 1833, with his tongue, perhaps, in his cheek, he wrote:

> Just before our departure [from Rome] we became acquainted with a fellow of Trinity College Cambridge [not otherwise identified] who, though he had not been in Italy much longer than ourselves, had attained an eminence so far beyond what we could even in thought aspire to that we gave the thing up in despair and retire upon the topos, that we don't enter into technicalities. Certainly those Cambridge men are wonderful fellows. . . . They know everything, examine everything and dogmatise about everything, they have paid particular attention to the geological structure of this place and the botany of that, and the agriculture of another, and they are antiquaries and artists and scholars and above all puff off one another. . . . Whewell's book and Sedgwick's lectures and Thirlwall's research and Hare's taste pop upon one at every turn.[1]

However, by 1866 an elderly Whewell saw his and that of the TC's world fast disappearing. He still retained a passion for riding, and this would be the death of him after being thrown from his horse a few miles outside of Cambridge that year. Any notion of a National Church central to the state and morality had also seemingly passed away.[2] This could be seen both intellectually and physically. For example, Whewell had been unable to prevent the rude arrival of "steam intellect" and the railway to Cambridge, but had been instrumental in ensuring the station was well away from the town and certainly out of view of Trinity.[3]

Not only was corrupt knowledge impregnating the university, but the landscape was also being adversely impacted. The railway was the iconic symbol of steam intellect and a utilitarian notion of progress. Many landed aristocrats feared the greater mobility of the lower ranks, geographically (and socially), would further erode their power and level society.[4] By the decade of his death, the onslaught of change and accompanying human-centric, anticlerical intelligence was now out of Whewell's control. His view of a liberal education had finally been eclipsed by a more vocational and utilitarian curriculum. In many ways his philosophy was now a relic of the past and seen as inappropriate to the new sciences, now branded by the materialism and industrialization that characterized modern Britain. His two closest friends, Hare and Jones, had died nearly a decade earlier, followed soon after by his wife, Cordelia. It was not long, however, before he had found comfort in the arms of another wealthy woman, Lady Everina E. Affleck. She was the recent widow of the rich banker Sir Gilbert Affleck, the eldest daughter of Francis Ellis, and sister of the former Trinity fellow Robert Leslie Ellis.

Whewell did not just make a great deal of money through lucrative marriages; he had also, despite his suspicion of the iron monsters puffing around the country, been a significant investor in railway companies. Whewell's divinity was seemingly compromised when it came to the sweet smell of profit. For example, in 1850 he invested £10,000 of his marriage trust from his union with Cordelia in the Lancaster and Carlisle Company.[5] Just over four years later he asked Jones to help him to decide which railway company to next invest in; should he put £15,000 in the North British or the Eastern Counties at 5 percent for five years?[6] Significant material change in the physical environment and the way it was experienced were paralleled in the intellectual terrain; speed of physical travel was also informing the speed of faith in pure human reason.

THE ORIGIN OF SPECIES

It was not just the demise of Euclidian geometry and its centrality in understanding God that was worrying Whewell. He confessed to Forbes that changes in science, such as Charles Darwin's theory of evolution (*The Origins of Species*, 1859) and contemporary debates on the age of the earth, were troubling him: "I cannot see without some regrets the clear definite line, which used to mark the commencement of the human period of the earth's history, made more obscure and doubtful." The Bible's claim that humans were only some six thousand years old and had been created in a period of six days now looked unsustainable. Whewell had spent all his life, along with the TC, trying to resolve the tenets of Protestant Christian religion, primarily that embodied in the Church of England, with that of knowledge, industrial development, and social stability. However, his whole edifice now seemed particularly vulnerable: "It is true that a reconciliation of the scientific with the religious views is still possible, but it is not so clear and striking as it was."[7] Already two years earlier in 1857 the well-known botanist and confidant to Darwin, Joseph Dalton Hooker, had sarcastically warned Whewell that we must "*draw upon our imaginations*, as to assume that species are created as such, & in one place only" (my italics).[8]

During this theological and intellectual turmoil Whewell championed his old-school friend, Richard Owen, and his demonstration of design in archetypal plans that existed only within the divine mind, most famously the vertebrate skeleton of a fish.[9] Whewell was part of an intellectual class, argues Adrian Desmond, that had "provided the channels through which Owen's pioneering morphology was to be promoted as the new conservative standard . . . [Owen] was now to reconstitute comparative anatomy along Romantic lines, to strip nature and society of their innate powers and return these to the Godhead."[10] Here was a Kantian/Schelling example of a Platonic transcendental archetype that went beyond simple empirical experience (for Whewell and Owen a divine—fundamental—idea).[11] Whewell was elated earlier when, as president of the Geological Society, it was his duty to award the society's prize medal to Owen—"a Lancaster man," he glowed, "the son of our former neighbour, Mrs. Owen—who is considered to be one of the first anatomists in Europe."[12] Owen, particularly, opposed transmutation and the idea that man originated from apes. He did not, however, fear works such as *Vestiges* since it still retained the gradual unrolling of a divine plan, uniformity, and a Supreme Intelligence.[13]

Whewell's support of Owen was entirely consistent, although not completely compatible, with his philosophy. For example, he similarly argued Gothic architecture grew from the introduction of the pointed arch. This feature lay at the center of the divine fundamental idea of verticality, which structured the whole church building. He said as much to an audience of British architects in 1862: "If we had only the lowest course of a clustered [Cathedral] pier before us, we could, in idea restore the building; just as the anatomist can, from the bones of the foot, restore the whole of the animal. There is an organic connexion and unity in the one case as in the other."[14]

Gothic architecture had general laws and an organic unity—it was just hard to explain. As Whewell slyly put it, "The ancient metaphysicians, in this respect wiser than the modern ones, made definition not the first, but the *last* stop in knowledge." Stemming from the pointed arch, an "internal principle of unity and harmony" commenced.[15] This unity of Gothic architecture, in contrast to the classical, represented perfect places for the worship of God. Likewise, following Owen, the design of animal skeletons from an ideal archetype was a divine "primal pattern" producing a blueprint for all vertebras.[16]

Whewell, like many others, did not think Darwin's subsequent theory was philosophically sound, since it did not explain the arrival of a rational and moral human being. To his death Whewell never doubted his own faith and accompanying epistemology. Man, especially an Englishman and in particular a Trinity man, was special, since unlike animals he developed in a specific sense, namely through a correct education—overcoming the Fall—progressively getting closer to his soul and thus the mind of God; to finding salvation and preparing for the future eternal world.

Archetypes stemmed from the divine mind and not nature as Darwin argued. The stakes were enormous. As Desmond and James Moore write, "The logic was stark—even if it was rarely spelled out. The day people accepted that nature and society would be torn apart, and would crash, the moral fabric of society would be torn apart, and civilized man would return to savagery."[17] Although, as Robert J. Richards has shown, Darwin retained an organic as opposed to mechanical view of nature, "and to identify God with nature, or at least to reanimate nature with the soul of a recently departed deity."[18]

Along with Whewell's college colleague, Sedgwick, he had two decades earlier tried to deal with the anonymous sensation of the subject of evolution, namely *Vestiges*. The idea was, in fact, not new and had been bandied about since the late eighteenth century—especially in France and Scotland. However, the notion of adaptation had been hijacked by political reformers and gained a new lease on life in the work of Darwin—a Christ's-educated gentleman-philosopher, unlike the anonymous and rudimentary speculations of the author of *Vestiges*.[19] Such a creed had the potential to dissolve the whole prevailing moral and social order. No wonder a version of his work was promoted with such gusto by the agnostic reformist "Bulldog" and destroyer of Owen, Thomas Henry Huxley. Darwin referred to Huxley as spreading the devil's gospel and, certainly he was considered a little rough, having not been broadly educated (Sydenham College—an anatomy school—and Charing Cross Hospital). He eventually replaced Owen as Hunterian Professor at the College of Surgeons and, with the appointment, the end of Owen's archetype along with its transcendental underpinning. God was removed from epistemology, and true faith was now invested in total human reason.[20]

Lyell had understood the adverse implications of evolutionary theory back in 1836. He told Whewell, had he "plainly stated as Herschel had done in his letter ['That mystery of mysteries, the replacement of extinct species by others']" to him "regarding the possibility of contradistinction to a miraculous process I should have raised a host of prejudices against me which are unfortunately opposed at every step to any philosopher who attempts to address the public on these mysterious subjects."[21] No wonder Whewell disliked him so much. Although Lyell took an inductive approach, he assumed "uniformity of law" without demonstrating it with the necessary evidence. Whewell branded him a "uniformitarian," unlike himself and others such as Sedgwick and the great Winchester College–educated and Oxford clerical geologist William Buckland, who all emphasized "catastrophism" via God's intervention. They would have shivered while reading Lyell's ditching of the Flood as well as discarding of Adam and Eve.[22] Whewell told Lyell there must have been an origin and his uniformitarianism was wrong.[23] In private, Lyell had already told Babbage how much he admired his argument concerning the governance of mechanistic laws:

> The argument of laws comes home to some of my geological speculations. No doubt some people would not like any reasoning which made miracles more reconcilable with possibilities in the ordinary course of the universe and its laws; but you do not write to please them. They are shocked at the idea of an eruption of a volcano being foreknown [like Babbage's Analytical Engine] which was to destroy Sodom and Gomorrah but your break in the sequence is more of a miracle than a volcanic eruption. *I think your estimate of the Creator's attributes much higher than theirs*, and that everyone may follow so far. (my italics)[24]

Lyell would soon find an ally in his geological uniformitarianism with Darwin's notion of gradualism in natural history. However, the St. John's botanist, onetime mentor to Darwin, and from 1837 rector of Hitchin, John Stevens Henslow, assured Whewell in 1847 about the transmutation of species—a query he had already been approached about by Sedgwick. His answer, he wrote, "must be the same—no Botanist, so far as I am aware, gives any credit to the tale." Indeed, he concluded, "In the present state of our knowledge I should as soon expect to hear that someone had seen a planet blaze out into a sun."[25] In private, Lyell's admirer, Darwin, disliked Whewell and the TC's world. He condemned anyone who found Whewell "profound, because he says length of days adapted to duration of sleep of man!!! Whole universe so adapted!!! & not man to planets—instance of arrogance" and an arrogant theology.[26]

Whewell was aware that science was now speculating that the earth and humankind were a great deal older than the Bible stated: however, he would die before the alarming discovery that human activity predated linguistic findings—although already in 1859 Neanderthal (a subspecies of early humans) remains had been found in a cave near Torquay, providing evidence that humans significantly predated biblical claims. On this revelation, as J. W. Burrow sums up, "philology could no longer plausibly claim to be the clue to the remotest period of man's history, and archaeology seemed to go far to justify the assumption of the evolutionary anthropologists, the purveyors of a rival, positivistic theory of history, that modern savages could be taken as a fair guide to the condition of the prehistoric ancestors of civilized man."[27] The instinctual bases of man was underlined by Darwin in his subsequent publications, including *The Descent of Man* (1871) and *The Expression of Emotions in Man and Animals* (1872).[28] Natural laws and, more controversially, contingency lay behind the evolution of all life, not the Deity.

It was Sedgwick who now hissed and puffed the most from Trinity over evolution. New species were a creation of God and nothing else—the fossil record, he claimed, shows this. He told his former student, Darwin, in 1859 that he "admired greatly" parts of the *Origin of Species*. However, he then scolded him: there were "parts I laughed at till my sides were almost sore; other parts I read with absolute sorrow, because I think them utterly false and greatly mischievous." Elsewhere he was even more hostile, and in an anonymous review of Darwin's book for the *Spectator* in March 1860, he snarled: "I cannot conclude without expressing my detestation of the theory . . . because it utterly repudiates final causes and therefore indicates a demoralised understanding of the part of its advocates."[29] Morality derived from God and therefore we were accountable to the Creator and not "self-evolving forces." For Sedgwick and the TC, such a materialistic explanation through natural selection meant "the end of civilization." For Darwin, it was the start of a new emancipating epoch in which the mind, morality, and religion could be explained via the evolution of the human species.[30]

WHEWELL'S ENLIGHTENMENT

During his final years, Whewell focused most of his time preaching to students in the Trinity Chapel. It was here that he was at his clearest and most authentic. It was here that his lifelong view toward religion and knowledge was most clearly expressed. His primary message was, as it had more or less been throughout his life, on the need to prepare in this temporary world for an eternal future in heaven. This was the true Enlightenment: "Set your affections on things above, not on things on the earth." Look up and not down, vertically and inductive, not deductive. He was consistent to the end on the function of a Trinity education. It was the task of the college to nurture the soul and ensure such a spirituality was carried on into whatever career the student next pursued. This defined, in a Kantian sense, the duties one was to pursue to obtain this objective. To follow "the conviction that you have, while here, been helped forward in that purpose from the things on earth to the things in heaven, which alone gives meaning and value to your career." We were all "servants of Jesus Christ"—our mediator with God—and "pilgrims on the journey to a better country."[31] It was thanks to Christ's crucifixion and subsequent resurrection that humans had the potential to access the will and mind of God.[32]

If you could save souls on the way, all the better. In a slightly earlier sermon celebrating the Holy Trinity and the establishment of Trinity College, he wrote that it was the function of education "to provide a religious and spiritual teaching for future ages," and underlined the limitations of human knowledge in comparison to "divine truth." The closer men got to God, the closer they got to each other.[33] As that important disciple of Coleridge and the TC, Maurice, had earlier written in his *Kingdom of Christ*, "Church Universal, *not built upon human inventions or human faith*, but upon the very nature of God himself" (my italics).[34]

In 1863 Whewell sermonized to Trinity students on the true constitution of knowledge. God had created man in his own image. The only knowledge of man's origin could come from within—"it must be divine, not from human sources." It came from a synthesis of divine innate ideas and experience. Whewell divided knowledge into two kinds: "There is a natural history of the world and a providential history of the world. The natural history of the world we collect from a survey of nature itself, and from stores of human records which past ages have accumulated. The providential history of the world we have given to us in our bibles." Providential history was of an altogether different type: "A revelation made by God to man, is surely not likely to be of such a kind or to have had an object. If the author and Governor of the world communicates any message to men, it must be something which concerns men's relation to Him." No doubt with Darwin in mind, he concluded that ultimate knowledge of the creation of man was an illusionary human quest: "The *origin*, in anything with which our thoughts can deal, is where we cannot clearly conceive or comprehend. The *origin* of matter, of life, how verily do our minds strive to form any conception of" (my italics).[35]

Whewell explained what was meant by the creation of man in God's image. This referred to "man's immaterial part—to his mind, his soul, his reason." Consequently man's soul—his reason—have "in it something of the return of God"; something divine that, in turn, made the soul eternal. This was a message that had consistently underpinned all of Whewell's history and philosophy of the sciences—as well as his moral philosophy. "Man too has his Logos—his mind—to which he looks at the things which were made. He uses this human mind in looking at and trying to understand the works of the Divine Mind"—a task made possible through the sacrifice of Christ. Whewell then posed a question to his flock: If we asked the great men of history "whether they have found, in their survey of men's knowledge of the world in which he lives, any mark, any fraction, which suggest to them the belief that there is in the soul of man, something Divine—that the human mind, when it contemplates the world, discloses some image of the Divine Mind which made the world, what do they say?" The answer was in the affirmative, namely a "divine image" could be detected, "however dim."[36] Science, Whewell sternly told Lyell, told us nothing of Creation.[37]

The detection of such a divine image could be traced, for instance, to the "earliest time of man's consciousness of his intellectual powers." Consider, for example, ancient Greek Euclidian geometry and its revealing of space. This "Divine principle in the human mind" was vital in disclosing this image. Space was a "fundamental idea" divulged via geometry (not algebraic analysis). "Are not the truths which we have to contemplate in our study of geometry of this character?" So too were the ideas of time and number made known through arithmetic—not to mention morality and duty. Such ideas, along with revelation, demonstrated for Whewell, that man was created and maintained in the image of God.[38]

However, Whewell acknowledged a problem: shortly after man was created in God's image, he fell from this state through sin, leading to a degenerate nature. This perversion of man's moral being made it appear that the soul was not in God's image. Nonetheless, Whewell preached, "The image upon his soul, according to which he ought to form his life, is still inside." It was just hidden from view by the dark shadow of sin. This led many men to see themselves as "opposite to the Divine Nature" and "hateful to Divine influences." It led them to faith in just their own perverted and immoral human reason. This was an artificial material and mechanistic view of the world. Utilitarian mechanistic projectors, like Babbage, were fallen men. Redemption could not come from man: "Because he was man and not God, he had no power to return himself when fallen: to purify himself from his foulness, to help himself in his weakness; to return himself to the image of God which he had lost." Redemption came through the sacrifice of Jesus Christ, the son of God, who died for our sins. Salvation subsequently came to those who love and believe in him. Christ's sacrifice made possible a knowledge of the will and knowledge of God.[39] Imitating human intelligence through mechanization told us nothing of true knowledge and simply took us further away from the Creator.

Not surprisingly, the centrality of Jesus Christ to salvation, knowledge, and the eternal life was a frequent theme with Whewell. It was only through Christ and his resurrection that our sins would be forgiven. A year before Whewell himself died he preached that one of the most striking passages in Scripture is the first verse in the opening chapter of the Gospel of John: "In the beginning was the word, and the word was with God." This, he continued, was not so dissimilar to the opening of the Old Testament: "In the beginning, says the book of Genesis, God created the Heaven and the Earth." The phrase "in the beginning" was common to both the New and Old Testaments. It was at this moment God started "his dealing with the world."[40]

The "word" refers to Jesus Christ, the son of God, "the inspiration of him who came into the true light. It was the Lord which was made flesh, and dwelt amongst us." The word "is used to express that endowment of Reason" and speech that man possesses, "but it is here used to signify the reason of God—some attribute of Him which we can best understand by looking at these attributes of man which this word indicates." Although there remained an "immeasurable difference, which we much suffer between what is human and what is divine." It was the ancient Greeks, primarily Plato and the Platonic school, which first clearly articulated this large gap between the human faculties and the "Divine Creative Mind." The only knowledge man could have of the "sensible world" came about by "determining some view of the Divine Ideas" that underpinned the Creation of God's world. "The Divine Reason, the Supreme Mind, the eternal Logos, must be conceived, if we are to conceive of the beginning at all, as the first offspring of the eternal Godhead; and the first step in the creation of the world. The final act of the Divine causation must have been the activity—the act—of that Divine Reason." All the gospels taught this, and this is what Whewell taught.[41] This was the true origin of species.

Only those with the "most reverent" and "religious spirit" could make deep discoveries of the visible world. And even here they would only be "glimpses of the scheme and fragments of the laws of God's creative power." Even more difficult was accessing those things that belong "to that other and higher part of man's nature by which he rises above the material world, to all that has reference to his moral, his religious, his spiritual nature." This world is also ruled by God: "But here we try in vain—the wisest of men try in vain—to discover laws, to penetrate mysteries." Yes, there was some form of communication between man and God, with the latter using revelation and the former prayer, but human language was no longer the word of God. Rather, "the light is no longer an intellectual light, it is a spiritual light. Man, as he approaches towards it and feels its power, no longer participates in the Divine idea, but in the Holy influence: he is *enlightened* only by being *sanctified*."[42] There was no human rational enlightenment only a holy one.

THIRLWALL'S TURN

By this stage Thirlwall was more concerned, ironically, about the danger of German biblical criticism and the future of the Established Church than the threat from evolutionary theory.[43] This ugly manifestation, for Thirlwall, emerged in a jointly written book entitled *Essays and Reviews* (1860). Biblical scholarship, primarily derived from the German School that Thirlwall had once so admired, had now greatly dissolved the historical standing of the gospels. It was not so much that no one had encountered German scholarship, but in this new volume it was given an established stamp of approval from predominantly Oxford and Cambridge scholars. The book marked a high-profile acceptance questioning the orthodoxy of traditional doctrine.[44] Bishop Blomfield greatly disliked Germanism and asked all his candidates for ordination, "I trust, sir, that you don't understand German?"[45]

This collection of essays, Thirlwall groaned, was a cold rational theological work, published four months after Darwin's *Origin of Species*, and was causing him far more anxiety. Ironically, an approach to biblical study that he had once so revered had now become anathema. The book was composed of six chapters edited by the English publisher and printer John William Parker, who, as we saw earlier, had published *Politics for the People*. Each chapter was written by a prominent theological figure: the Balliol graduate and headmaster of Rugby School Frederick Temple, the Oxford professor of geometry Baden Powell, the Oxford St. John's tutor Henry Bristow Wilson, the former Oriel tutor and future rector of Lincoln Mark Pattison, the Balliol fellow Benjamin Jowett, and two dons from Cambridge, Rowland Williams—now professor of Hebrew at St. David's College, Lampeter, and the only layman among the essayists—and the Egyptologist and lawyer Charles Wycliffe Goodwin.[46]

Bishop Thirlwall, by now much more protective of and fearful for the Established Church, joined Bishop Samuel Wilberforce to condemn the work for heresy. One of the most polemical, critical, and sarcastic responses to the *Essays and Reviews* came from Christopher Wordsworth's son, the Trinity graduate Canon Christopher Wordsworth.[47] The arguments of the liberal essayists culminated, Thirlwall concluded, in the entire eradication of Christian faith. There was nothing left of the revelation of supernatural truth, and all that remained was some vague emphasis on an ethical and moral system; German biblical criticism, as Rose long ago predicted, had cleansed Christianity of everything. The whole of Christian religion had been rationally stripped bare by human reason and left with no content—as such it was also morally vacant. Thirlwall succinctly spelled out the grave result: "It will no longer be a religion, and will not yet have become a philosophy. No longer a religion, because it will contain nothing which is not supposed to have been originally derived from the processes of unassisted human reason. Nor yet a philosophy, because it will retain many traditional elements, and will still appeal to authority in matter on which reason claims a supremacy, which, at the present stage of the education of the world, can no longer be questioned."[48] This had been the TC fear all along, and it had now become institutionalized and made manifest.[49] Thirlwall's favorite onetime student, Maurice, scolded the authors of the *Essays and Reviews*, telling them that if they had "more fully confessed that God had revealed Himself to us, and that He is guiding us by His spirit out of the opinions of men and ages, into the Truth which belongs to all ages, they would be more consistent Churchmen."[50]

The collection of chapters in *Essay and Reviews* dominated the concerns of the Anglican clergy far more than *Origin of Species*.[51] What made things particularly bad was the fact that several of the authors were ordained and teaching the future clergy at Oxford and Cambridge. If they held such views, many staunch Anglicans argued, they should leave the Established Church.[52] This sentiment could also be made by sympathizers. For example, the Wadham tutor, ex-Christian, and now on his way to becoming a positivist, Frederic Harrison, believed that the essayists should have resigned—this being the logical conclusion to their argument. He wrote in the *Westminster Review* that year, "What becomes of the Christian scheme when the origin of man is handed over to Mr. Darwin; and Adam and Eve take their seats beside Deucalion and Pyrrha?" Darwin's work may not yet have made the impact it would, but it certainly got absorbed, at some level, within the fury over *Essays and Reviews*.[53]

Of all the essays published, Thirlwall thought Powell's the most heretical and dangerous. There was no conflict between science and religion, claimed the Oxford don, since they had no common ground. Nor did miracles exist as external historical facts. Powell severed historical revelation from Christianity and left, just as Thirlwall stated above, an indefinable spirituality—not a religion. This, Thirlwall believed, was a short step toward atheism. This indefinable spirituality, however, as we will see in the Epilogue, would form the future bedrock, for many, of Christianity.

Wilberforce hissed that the book was "a lawless rejection of all authority" and a "daring claim for the unassisted human intellect to be able to discover, measure, and explain all things." Scripture was simply to be accepted and not judged. However, Babbage and the world of his ilk had somehow become dominant. Thirlwall, now holding completely different views to his earlier stance at Cambridge, strongly advocated the need to prosecute the book for "prolate heresy." It should also be remembered that in his revised chapter against miracles as evidence, Powell inserted a reference to Darwin's *Origin of Species*, predicting the work would "soon bring about an entire revolution of opinion in favour of the grand principle of the self-evolving powers of nature." This no doubt infuriated Thirlwall further. However, Thirlwall was subsequently reminded by admirers of the book about his previous translation of Schleiermacher and stand toward allowing religious Dissenters into Cambridge; the ghost of earlier youthful enthusiasm had come back to bite him hard. Thirlwall subsequently diluted his reaction to *Essay and Reviews*.[54]

It was left to others to condemn the links being made between Darwin's book and *Essays and Reviews*. For example, the newly appointed rector of St. Nicholas, Worcester, Charles Bullock, wrote, "It is now universally admitted, that the pretentious theory of 'The Vestiges of Creation' is opposed to the teachings of true science. We hear no more of progressive development . . . from the 'monkey' to the 'man' . . . Darwin's more recent theory of 'The Origins of Species' is propounded with greater ability, but is not better sustained by facts." The governor of Madras and brother of Archbishop Denison, Sir William Denison, thought he would wade in too with an attack in 1862: "Mr. Baden Powell would accept the theories of LAMARK and DARWIN as scientific facts, which others would consider as mere visionary hypothesis." The lay leader of the evangelicals, the 7th Earl of Shaftesbury, lashed out with a blistering attack: "Only let the doctrine of the development of species be taken up by dignitaries or men of position . . . and put forth to the public, and hundreds will be found to rejoice in the discovery that they are sprung from a fungus, and that as soon as possible to a fungus they will return (Laughter)." His point, however, was deadly serious. Scripture was as suited to our times as much as it always had been.[55] Although peripheral, the argument over natural selection and the origin of species still provoked controversy among scientists.[56]

Essays and Reviews marked a landmark juncture in the relationship between the Church and the centrality of orthodox doctrine at the center of a Cambridge and Oxford education.[57] The Bible should be treated like any other book and the findings of science related and applied to it. Revelation should not be left to the Church but to an individual conscience. A person purporting the authority of Scripture should thus be equipped in the necessary critical tools to interpret the work, and as such, it should not be left to a pastoral cleric. This was a shift away from the ecclesiastical and to the trained intellectual; such people should be free to interrogate the Bible. Human freedom, energy, and creativity were crucial to progress. Science should be embraced, and the destructive wall of monopolistic tradition knocked down to ensure such progress. As such, universities had to be reformed to accommodate this. Perhaps more than anyone else in *Essays and Reviews*, Jowett was at the vanguard spearheading such reforms. Shortly after he was made master of Balliol in 1871, the need to subscribe to religious tests at Oxford, Cambridge, and Durham was abolished.[58] If German biblical scholarship had been seen as heretical early in the century, it was now seen as crucial. In addition to the Germanic turn in *Essays and Reviews*, the entry for "The Bible" in the *Encyclopaedia Britannica* in 1878 was almost entirely drawn from German divines, whereas in the earlier seventh (1830–1842) and eighth (1852–1860) editions there was next to no German scholarship.[59]

It was not, however, all doom and gloom. In 1869 Thirlwall was given an honorary fellowship at his old college—some thirty-five years after his semi-expulsion over the Dissenting affair in 1834. The invitation came from his old pupil, the Cambridge professor of Greek and now master of Trinity William Hepworth Thompson, who took over from Whewell in 1866.[60] Nonetheless, much of what the TC had tried to preserve had now been overturned. Human reason alone, via science, had now become the predominant glue of understanding and the basis for a university education. Thirlwall died in 1875.

EPILOGUE

THE DEMISE OF GOD AND THE RISE OF MACHINE INTELLIGENCE

The world for the TC was divine and alive. It was organic and needed nutrients provided by an active intervening God that depended on fertile faith; life was not mechanical with a nonexistent or nonactive Creator. However, for the last living member of the TC, human logic had displaced God. Thirlwall's melancholy conclusion over the damaged credibility and foundation of Christian faith, and thence too the Established Church, was not accepted by all. For some a reconciliation over Christian doctrine and science could be found—a key aspect, after all, of the *Essays and Reviews*.

A group concentrated at Balliol, and Oxford in general, employed a philosophical idealism that dismissed dogmatic doctrine and underlined a (German) spiritualism—very similar, ironically, to that expounded decades ago by Thirlwall and Hare. The harbingers were Benjamin Jowett, one of the instigators of *Essays and Reviews*, and his student Thomas Hill Green. Vulnerable and controversial beliefs over such things as original sin, atonement, and miracles were irrelevant. Rather, God had become flesh through Jesus Christ, and thus the emphasis should be on the Incarnation; his body was the Church and should articulate the comradeship of man. Salvation switched to serving mankind.[1] This made it possible to have faith in human reason since humankind was God.

Green later wrote in his *Prolegomena to Ethics*, published a year after he died (1884), "The good which a man seeks for himself is not a succession of pleasures, but objects which, when realised, are permanent contributions to a social good which thus satisfies the permanent self."[2] Self-development cashed out as the most appropriate approach to fulfilling the best for the community; this was the key to creating the ideal environment for men and women to develop their intellect and morality. If science and historical criticism were properly understood it would restore and bolster Christian faith.[3]

This idealism, like the TC's earlier, was also an attempt to confront the cold, heartless, mechanical outlook of utilitarianism. The quest was for inward perfection. This was a metaphysical quest, as opposed to material, resulting in the common good—a collective spiritual unity.[4] The universal human essence, the TC believed, was embodied in the organic evolution of our consciousness, institutions, constitution, and terrestrial social hierarchy. Thus it was important to preserve and encourage them in the Burkean sense. Likewise, an eternal human consciousness described all men—this was a spiritual impulse from God that informed man's morality and action.[5] In many ways the ethical economy the TC expounded was transmuted into the morality of the idealists and subsequently the ethical socialism of reformers such as Richard H. Tawney; all targeted utilitarianism and individualism. Of course there were fundamental differences. For example, the TC was a conservative Tory Anglican movement that emphasized the future world and Scripture. This in turn was replaced by an idealist—more socialist—emphasis on the present and an incarnate God. God had manifested as us, via Jesus Christ, and this glued everyone together; individuals were spiritually bound.[6]

The working-class solidarity Tawney later found characterizing the northwest of England was not so different from the traditional ruling-class solidarity the TC had earlier sought to defend. At its heart were Christian belief, knowledge, and morality—all held together, for the latter, by a unified National Church. Tawney was educated at Rugby and Balliol at the same time as his friend William Temple, the future archbishop of Canterbury. They were joined at Balliol by the Charterhouse-educated William Beveridge. Where the idealism of the TC was to nurture a faith in Scripture and salvation to ensure the correct individual morality and intelligence, the new idealism was less a dogmatic religion and more a morality and idea of citizenship. As Melvin Richter puts it, "Religion, so conceived, is independent of miracles and of all statements presented as historical or scientific within the Bible."[7]

Where the recent research of Tim Rogan places the primary precursors to Tawney within Britain in the works of Thomas Carlyle, John Ruskin, and William Morris, this book has shown that aspects, despite their conservative outlook, can be found in much of the ponderings of the TC.[8] All despised the piercing of a moral solidarity by the mechanistic individualism of a new capitalist creed supported by a philosophy of utilitarianism. For the TC, and shortly later the Balliol idealists, morality had nothing to do with the quantitative measurement of pain and pleasure.[9] However, unlike the new idealists, Whewell and the TC were not interested in the growing emphasis on dramatic terrestrial material inequality.[10] By contrast, for the former this was simply an inexcusable argument for not doing one's moral duty here on earth. This was a major difference. For the TC it was all about preparing for the next world—personal salvation—whereas the idealists looked to improve the here and now; there was no detachment between man and God. Justice was to be done now and not in some apparent next world through community improvement. Both, however, underlined the need to reach a higher state and morality.

Spearheading the quest to advance the lives of common folk was the Rugby- and Balliol-educated Green. He remained at his college and became Tawney's teacher. Green was a supporter of the university extension movement, expanding the franchise, securing the community, equal opportunity, improving working class conditions, widening trade unions, and the Co-Op Movement. One of his leading followers at Balliol was the social reformer and popularizer of the term *Industrial Revolution*, Arnold Toynbee. The objective of Green and the Balliol idealists was a consciousness of the common good made possible by a state and society where individual self-realization could be nurtured. If society did not exist, neither did the individual. For these Balliol idealists religion was something practiced in everyday life and not learned through Scripture. Human reason was God given and not simply human. Like the TC, and unlike the Oriels, Christian morality was not severed from the economy; indeed, it should define an ethical economy (as opposed to political).[11]

The inheritors of this idealist philosophy—led by Toynbee and the Eton- and Balliol-educated Henry Scott Holland—strongly opposed, as did the TC, political economists who expounded universal laws. As Holland warned in *Our Neighbours* (1911), "If you leave forces to work blindly, you are sure to back the strong, at cruel cost to the weak." Moreover, "*If you think that human power can ever work mechanically, like material energies*, you have misinterpreted human nature, and betrayed its honour. Your product will be something inhuman" (my italics).[12] Likewise, Toynbee earlier despaired in his *Lectures on the Industrial Revolution* (1884), "A law of industrial development has designated him [man], by division of labour, *from a man into a machine*" (my italics). Laissez-faire economics needed to be replaced by Christian moralized competition, not deduction, crude empiricism, and abstract analysis.[13] These predominantly Balliol scholars assimilated aspects of the TCs idealism and represent the ethical socialist bridge to Tawney and later historians such as the Balliol scholar Christopher Hill and Cambridge Corpus Christi graduate E. P. Thompson.

The TC offered a nineteenth-century lens on a then-prominent, but now largely forgotten, critique of capitalism's authorative axioms, namely deductive political economy. Implicit to this was an alternative notion of rationality to the view that eventually won. In the new commercial and utilitarian environment, God had been replaced by human reason and an increasingly deterministic view that reduced everything to economic measurement revolving around competitive self-interest. The market had literally replaced God. The potential was now open to financialize every aspect of life.[14] For the TC, the Cambridge curriculum, via its emphasis on coaching and mathematical analysis, had become nothing but a competition characterized by a false and narrow focus; the human faculties were not being cultivated and students were not receiving an education. Faith was being erased by a false human logic. For all of Whewell's detailed work on the history and philosophy of science, his aim was simple: it was to show that there was no systematic scientific method of discovery. Behind all known knowledge was a divine fundamental idea available only to those who held the right morality born from the correct religious belief. To understand this idea required a contingent form of understanding, that is, belief in the Bible and Euclidian geometry. Faith always will, or should, win over mere human reason.

Green and the Balliol idealists changed this by emphasizing Incarnation and placing God, via Jesus Christ, as man. Guilt at stark social inequality should be reconciled via Christian values and a spirit of brotherhood. As Toynbee preached to a private gathering at Balliol in 1879: "After faith comes *knowledge* - how shall we know God? How detain this ideal that hovers like winged light within the mind? To know God man must seek to become God - life is the ceaseless endeavour to become like God; to enact God in our own souls and in the world; and though men must needs fail, failure here is the only success."[15] This was the new Anglican answer to utilitarianism and would go on to inform the work of Tawney.[16] In *The Acquisitive Society* (1921), Tawney condemned the mechanistic philosophy that now dominated humanity. Instead, he emphasized the Incarnation and bemoaned the banishment of providence that had now seemingly become a "spectator of the curious machine" created by man. This destructive machine, he warned, had been created and set in motion with no one in control.[17] This false society was dying because it had no morality.[18] As Toynbee summed up, "*A logical artifice became the accepted picture of the real world*," with "Ricardo's brilliant deductions" destroying "observation" and, with it, morality (my italics). In this sense the Balliol idealists and subsequent moral economists were on the same page as the TC.[19]

This view, however, was very much against the tide. In a growing material and industrial context, Cambridge had been forced to rebrand itself as producing perceived useful citizens for a new capitalist era. Indeed, it is extraordinary how an institution such as Cambridge, after some five hundred years of being primarily a body for Christian seminaries, so quickly transformed into an organization teaching the sciences, social sciences, and humanities. Karl Polanyi and Thompson, in some sense the torchbearers to Tawney, continued to emphasize "human personality," albeit via a secular alternative, namely a humanist Marxism. Christian idealism was now, as Thirlwall feared, stripped of its faith and had become a socialist doctrine. Christian values mutated into an overt critique of and an attempt at overcoming capitalism but this time without Christianity and instead a new faith termed Marxism; the prophet was not God through Jesus Christ but the terrestrial political philosopher, Karl Marx. Both Polanyi and Thompson, however, would soon become disillusioned and seek an alternative critique of utilitarian capitalism.[20]

Thompson was educated at the private Dragon School in Oxford, followed by the Methodist private school Kingswood and finally Corpus Christi, Cambridge. Like the TC and Balliol idealists he underlined the importance of customs, tradition, and a "moral economy."[21] All provided an ethical critique of capitalism. As with the TC earlier, Tawney, Thompson, and Polanyi harked back to the medieval period and bemoaned the shattering of an organic community that, they believed, characterized that period. Aspects of medieval morality, via tradition and customs, remained embedded in eighteenth-century England.[22] For them, as it was for the TC and Balliol idealists, it was men like Ricardo and Bentham that spelled its end and replacement with a human-centric capitalist rationality. Modern science emerged in the second half of the nineteenth century appearing to support this move. In the process, as Peter Harrison puts it, "the older moral and theological unifying principles" were erased. "Science" became "fundamentally a human activity, and the major player had become the scientist. Prior to this, nature itself was the focus and the chief player, albeit one who transcended physical phenomena, was God."[23] The *scientist*, a word Whewell coined, had ironically come to represent those at the apex of this new world.

Another important inheritor of the TC's perspective was William Cunningham, yet another product, in some sense, of Trinity. He studied in Edinburgh before going to the University of Tübingen, then a hotbed for the German historical approach. Cunningham subsequently returned to Britain and decided on a career in the Church of England. He then commenced the study of the moral sciences under Henry Sedgwick at Caius, before migrating to Trinity and gaining a first in the subject with his friend, the Eton-educated Frederic William Maitland.[24] The latter went on to win the Whewell Scholarship in International Law. In theological matters Cunningham was inspired by the work of Hare and especially Thirlwall's favorite student, Maurice.

Like Tawney, it was while teaching in the north and northwest that Cunningham witnessed the troubled lives of the industrial proletariat. Thence commenced his interest, like Toynbee's earlier, in economic history and critique of laissez-faire economics. Here he applied the same inductive approach as Jones and Whewell, with an emphasis on economic understanding being dependent on time and place.[25] In 1887 he was appointed vicar of the large Cambridgeshire parish of Great St. Mary's, which he combined with a lectureship and chaplaincy at Trinity. His emphasis on imperialism and national economic regulation put him at loggerheads with the dominant Cambridge economist, Alfred Marshall. Cunningham despised Marshall's deductive abstract approach to economics as, like the TC had argued concerning Ricardo, having no historical foundations. Cunningham, like Peacock earlier, was finally made archdeacon of Ely in 1907 until he died in 1919.[26]

Cunningham underlined the importance of tradition, custom, and community. In this sense he shared significant overlaps with Polanyi and Thompson; of course the latter had no positive place for religion. Instead, as we have seen, they strove to reveal a secular and nonreligious set of customs and traditions. Nonetheless, for all of them money-making self-interest had dissolved the passions and now set the moral tone.[27] Intrinsic to this development was the rise of a commercial and indeed utilitarian middle class that the TC had so earlier despised. Their argument, that allowing everyone to pursue one's self-interest would ultimately lead to the greater good, was for all of them dangerous nonsense. The obsession of reducing everything to rational calculation in the pursuit of economic advantage drove them all mad. Placing the individual above God or the community was plainly wrong and, they argued, historically mistaken. However, the quest to reduce everything to some tangible measurement had to spell the demise of some incalculable metaphysical quality such as divine belief, fundamental ideas, or, in the case of the Balliol idealists, a unifying spirit of humankind.

The inheritors of aspects of the TC, the Balliol idealists and twentieth-century moral economists, were first battered by the liberal classical economists before finally being defeated by the rise of what we can loosely refer to as neoliberalism. However, even here you could be a staunch free marketer and still dislike material calculation and utilitarianism, as demonstrated especially in the work of Friedrich A. Hayek. This important twentieth-century scholar, a younger contemporary and onetime colleague of Tawney at the London School of Economics, is interesting for several reasons. He believed—in stark contrast to the TC, Balliol idealists, and twentieth-century moral economists—that the rise of the free market was central to civilization and a new morality; only successful self-interested humans gained ascendancy via group selection. However, he was also very Burkean, and indeed Whewellian, in arguing that old truths remained vital in the present.[28] Indeed, self-interest and human logic were subsumed to group selection. Likewise, Hayek was no fan of pure human reason. He boiled the rise of this human-centric faith completely down to reasons born of the French Revolution—somewhat akin to the TC. Nor did Hayek like divine design arguments, and instead emphasized spontaneous individualism that evolved through no recognizable consciousness.[29] Ultimately, of course, Hayek's conclusion was very different from those above: it was those humans who survived and prospered the most via a commercial reason, not those with a soul or bound to some regulated community, who mattered the most; for him this was cultural evolutionism. The market was the basis of knowledge—although you could never have a total understanding of either. To ensure the market worked to its optimum, it needed to be defended by international law and global institutions. This would protect it from national democracies born from the ashes of empire.[30]

Hayek harked back to the arguments of commentators the TC abhorred—such as the Anglo-Dutch maverick physician Bernard Mandeville and subsequent eighteenth-century British empiricists like David Hume—as the originators of his perspective. It had been, after all, Mandeville who emphasized the self-interest of an individual bee as equivalent to an individual human. The former pursuing its interest for the common good of the hive and the latter for the wealth of society, a point later pursued in the nineteenth-century political economy of Whately.[31] Knowledge for Hayek, unlike the TC, was not divinely innate but born of human capitalist cultural evolution.[32] It was perversely those negative traits associated with the Fall that now, ironically, lay at the heart of the modern capitalist world and thus should be celebrated.[33] For Hayek it was capitalism that bound humans together. All of them, however, shared a view on the limits of individual human reason.

Like Whewell, Hayek was Kantian in the sense that empirical observations made no sense until they were arranged by the mind. However, unlike Whewell or Kant, these evolved via capitalist cultural evolution and depended on collective information (not so different from Whately). The world economy was for Hayek, as Quinn Slobodian writes, "an enormous information processor beyond the capacity of the human mind to either manufacture or comprehend." Humans derived rules via an evolutionary process in which the lasting rules remained secure and the others disappeared.[34] Hayek wrote, "A certain part at least we know at any moment about the external world is therefore not learnt by sensory experience, but is rather implicit in the means through which we can obtain such experience."[35] This form of cultural evolution had found its true state in the tacit rules and accompanying morality of the free market; a conceived divine spirit unifying humankind was replaced by a conceived notion of a free market. This was what, for Hayek, defined modern civilization. A large and popular historiography has grown up around this perspective.[36] This was the antithesis of the nineteenth-century TC, Balliol idealists, and twentieth-century moral economists. For Hayek and his followers, it was not faith in God, or some sort of spiritual empathy that defined our morality but rather the human species's population growth, progressive cultural knowledge, and increasing efficiency via capitalism.[37]

The irony is that the social engineering that the TC feared so much at the hands of positivist utilitarians, and that Hayek believed led to totalitarianism, has manifested in the engineering mentality of the modern "quantitative" neoliberalism that defines the Western world today (not that of Hayek and the Austrian School).[38] What had once emerged as an emancipating ideology to smash the clutches of the nineteenth-century National Church and accompanying gentlemanly elite, had or has evolved to produce its own elitist form of social authority: simply a new sovereign monopoly in a different dress. The TC's notion of divine genius has been replaced by an elite, calculating cultural entrepreneur.[39] The pursuit of knowledge is about utility and the exploitation of the natural world. Morality has been reduced to a visible calculation based on the best outcome for the greatest number and not some mystical inner morality; Benthamite utilitarianism has been totally financialized and penetrates everything, including, of course, morality and education. One dogma replaced by another, and one elite replaced by another, although, perhaps predictably, the institutions educating them remain the same.[40]

The mechanistic individualism the TC so feared has not just risen to intellectual ascendancy but transformed the environment and the way we are ruled. What started off as a metaphor during the seventeenth century to describe an ideal political structure became a manifest possibility during the nineteenth century. This, as Jon Agar has shown, triggered the move in Britain to government as a machine and, indeed, has played a key role in the history of the computer.[41] For utilitarians the machine as metaphor was used as offering a neutral alternative to a corrupt, hierarchical, Church-based form of rule. For the TC a leading propagator of this reformist movement was Babbage. The latter knew he had to first prove its objectivity by demonstrating that such mechanization was based on human thinking and an absent God or, perhaps more accurately, that humans were God. On one hand you have an impetus to mechanical administrative developments, and on the other the creation of a formalized logic purporting to represent the reason of the mind and therefore God. The result is the establishment of an effective process of organizing thought and people to a supposed natural and thus neutral mechanism. This is the world we live in today.[42] As one disciple and historian of artificial intelligence (AI) recently put it, "the ghost in the machine is code," and we can seemingly build ourselves. However, as he also concludes, and I concur, "What a historical irony it would be if the intelligent machines that we created to be like us end up transforming us to become like them."[43] The Creator is re-created. Although as a final caveat, a leading critique of AI by many commentators claim that since human language is a distinctly social action, it cannot be properly reproduced by a computer—not so dissimilar, but for very different reasons, to the argument of the TC.[44]

NOTES

PROLOGUE

1. Peter Harrison, *The Fall of Man and the Foundations of Science* (Cambridge: Cambridge University Press, 2007) and *The Territories of Science and Religion* (Chicago: University of Chicago Press, 2015, 2017); Larry Siedentop, *Inventing the Individual: The Origins of Western Liberalism* (2014, St. Ives: Penguin Press, 2015); Stefano Franchi and Francesco Bianchini, eds., *The Search for a Theory of Cognition* (Amsterdam: Rodopi, 2011); George Zarkadakis, *In Our Own Image: Savior or Destroyer? The History and Future of Artificial Intelligence* (2015; repr., New York: Pegasus Books, 2017).

2. Perhaps the most well-known atheist exponent of Darwinian natural selection is Richard Dawkins; see especially his *The Selfish Gene* (Oxford: Oxford University Press, 1976) and *The God Delusion* (2006, Suffolk: Transworld, 2016). For a converse argument by the geneticist and head of the Human Genome Project, see Francis Collins, *The Language of God: A Scientist Presents Evidence for Belief* (London: Simon & Schuster, 2007).

3. This is a purely materialistic reading of Charles Darwin that omits much of the German Romantic aspect that informed his belief and development of evolutionary theory; see Robert J. Richards, *The Romantic Conception of Life: Science and Philosophy in the Age of Goethe* (Chicago: University of Chicago Press, 2002), esp. chap. 14. This black-and-white demarcation between Christianity and liberal secularism is, according to Larry Siedentop, a complete distortion of their entwined history; see his *Inventing the Individual.* The history of the Fall and its historical impact on the limits to human knowledge is best told in Harrison, *Fall of Man*. For Saint Augustine, the Fall of Adam meant both a moral and epistemological loss. Consequently, humans frequently conflated falsity with truth. There was still an image of God but now just a dim reflection that demanded God's intervention, which in turn depended on one's spiritual purity. For some, too, the limits of the senses could be improved via instruments such as a microscope or telescope. This surely was not missed in Charles Babbage's attempt to build a form of human intelligence in a machine; see Simon Schaffer,

"Babbage's Intelligence: Calculating Engines and the Factory System," *Critical Inquiry* 21, no. 1 (Autumn 1994); and William J. Ashworth, "Memory, Efficiency, and Symbolic Analysis: Charles Babbage, John Herschel, and the Industrial Mind," *Isis* 87, no. 4 (1996).

4. John Bellamy Foster, Brett Clark, and Richard York trace the history of this clash from antiquity to the twentieth-century intelligent design movement. Although their account is primarily a polemical critique of the latter, it is clear and provides a useful overview of the leading works and arguments; see their *Critique of Intelligent Design: Materialism versus Creationism from Antiquity to the Present* (New York: Monthly Review Press, 2008). For the Gallup poll, see Art Swift, "In U.S., Belief in Creationist View of Humans at New Low," Gallup, May 3–7, 2017, https://news.gallup.com/poll/210956/belief-creationist-view-humans-new-low.aspx.

5. Stephen Jay Gould, *Rocks of Ages: Science and Religion in the Fullness of Life* (1999; repr., London: Ballantine Books, 2001), 3–10. For a recent defense of Intelligent Design that does not partition science and religion into different domains, see, for example, William A. Dembski, *Intelligent Design: The Bridge between Science and Theology* (Downers Grove, IL: InterVarsity Press, 1999).

6. This was, in fact, a centuries-old debate; see Harrison, *Territories of Science*. Harrison clearly demonstrates the historical poverty within current popularist views of science and their sharp demarcation with religion. A related argument is also made in Siedentop's *Inventing the Individual*.

7. Theodore M. Porter, *Trust in Numbers: The Pursuit of Objectivity in Science and Public Life* (Princeton, NJ: Princeton University Press, 1996); Harrison, *Territories of Science*, 188–92; Siedentop, *Inventing the Individual*, esp. 349–63. For its latest manifestation, see William Davies, *The Limits of Neoliberalism: Authority, Sovereignty, and the Logic of Competition* (2014; repr., London: Sage, 2016).

8. Albert O. Hirschman, *The Passions and the Interests: Political Arguments for Capitalism before Its Triumph* (1977; repr., Princeton, NJ: Princeton University Press, 2013). Perhaps the ultimate severing of the passions was in the work of the Reverend Thomas Malthus who placed the biological imperative of reproduction at the heart of human existence. Gareth Stedman Jones writes, "Once this conception was implanted at the heart of political economy, the core of economics was henceforth situated in the realm of nature"; Jones, *An End to Poverty?* (London: Profile Books, 2004), 6. A prominent twentieth-century economic commentator who, conversely, equated the rise of capitalism with the rise of civilization was Friedrich A. Hayek. However, his focus was on cultural evolution and the politics of population growth. Not surprisingly Malthus is not mentioned in his analysis. For an excellent overview, see, for example, Naomi Beck, *Hayek and the Evolution of Capitalism* (Chicago: University of Chicago Press, 2018).

9. For a recent attempt to rehabilitate the place of God and the imagination in acquiring real knowledge—drawing particularly on the work of Samuel

Coleridge and William Wordsworth—see especially Douglas Hedley, *Living Forms of the Imagination* (New York: T & T Clark International, 2008).

10. Otto Mayr, *Authority, Liberty and Automatic Machinery in Early Modern Europe* (Baltimore: John Hopkins University Press, 1986), 135, see esp. chap. 6, 126–30. For more recent iterations of this argument, see, for example, Deirdre N. McCloskey, *Bourgeois Dignity: Why Economics Can't Explain the Modern World* (Chicago: University of Chicago Press, 2010); Joyce Appleby, *The Relentless Revolution: A History of Capitalism* (New York: W. W. Norton, 2010); Joel Mokyr, *The Enlightened Economy: Britain and the Industrial Revolution, 1700–1850* (New Haven, CT: Yale University Press, 2011), and Mokyr, *A Culture of Growth: The Origins of the Modern Economy* (Princeton, NJ: Princeton University Press, 2018).

INTRODUCTION

Epigraph: William Whewell, *On the Philosophy of Discovery: Chapters Historical and Critical* (London, 1860), vii. He later reiterates the same point: "The world is constituted according to the Ideas of the Divine Mind, and that the human mind apprehends the inward and most fundamental relations of the universe by sharing in some measure of these same ideas" (364). This argument is intrinsic to his *History of the Inductive Sciences*, 3rd ed., 3 vols. (London, [1837] 1857), and *The Philosophy of the Inductive Sciences Founded upon Their History*, 2 vols. (Cambridge, 1840).

1. W. G. Clark, "William Whewell: In Memoriam," *Macmillan's Magazine*, 1866, 545.

2. For Whewell's most blunt and vibrant published expression of this argument, see *The Elements of Morality Including Polity*, 2nd ed. (Cambridge, [1841] 1854), 27–31, and *Of the Plurality of Worlds*, 4th ed. (London, [1853] 1855), 376–79. It is also alluded to in his *History of the Inductive Sciences*, 1:6. The full force of this perspective was given in his mainly unpublished sermons preserved in the Whewell papers at Trinity.

3. See especially Whewell, *Philosophy of the Inductive Sciences*, 1:xvii–xxxvi ("Aphorisms Concerning Ideas"), 26–75. For Lutheranism, see Harrison, *Fall of Man*.

4. Whewell, *Philosophy of the Inductive Sciences*, 1:26–27.

5. Richards, *Romantic Conception of Life*, 70–71; and Willem R. de Jong, "Kant's Theory of Geometrical Reasoning and the Analytic-Synthetic Distinction: On Hintikka's Interpretation of Kant's Philosophy of Mathematics," *Studies in History and Philosophy of Science* 28, no. 1 (1997): 143–57.

6. Hugh Evan Hopkins, *Charles Simeon of Cambridge* (London: Hodder & Stoughton, 1977), 18.

7. N. Merrill Distad, *Guessing at Truth: The Life of Julius Hare 1795–1855* (Shepherdstown, WV: Patmos Press, 1979), 6–7; and G. F. McFarland, "The

Early Literary Career of Julius Charles Hare," *Bulletin of the John Rylands Library* 46, no. 1 (September 1963): 43–44. Francis Hare-Naylor was best known as the author of a *History of Helvetic Republics*, 2nd enlarged ed., 4 vols. (London, [1801] 1809), and *Civil and Military History of Germany, from the Landing of Gustavus to the Treaty of Westphalia*, 2 vols. (London, 1816). Julius Hare's passion for Greek history and languages was probably further fueled by his uncle on his mother's side, the well-known Asian and Sanskrit scholar Sir William Jones. Georgina Shipley was also the daughter of William Shipley, who established the Society for the Encouragement of Arts, Manufacture and Commerce in 1754; see Sara Gray, *The Dictionary of British Women Artists* (Cambridge: Lutterworth Press, 2009), 240.

8. I am grateful to Chris Stray for this reference to Clotilde Tambroni. The importance of Rome to the Hareses and the nineteenth-century imagination is told in Robert O. Preyer, "Breaking Out: The English Assimilation of Continental Thought in Nineteenth-Century Rome," *Browning Institute Studies* 12 (1984).

9. From 1808 Clarke was Cambridge Professor of Mineralogy, succeeded by John Stevens Henslow in 1822 and Whewell in 1828.

10. See, for example, Melvin Richter, *The Politics of Conscience: T. H. Green and his Age* (1964; repr., Bristol: Thoemmes Continuum, 1996); D. M. Mackinnon, "Some Aspects of the Treatment of Christianity by the British Idealists," *Religious Studies* 20, no. 1 (1984); Matt Carter, *T. H. Green and the Development of Ethical Socialism* (Exeter, UK: Imprint Academic, 2003); Jonathan Sheehan, *The Enlightenment Bible: Translation, Scholarship, Culture* (Princeton, NJ: Princeton University Press, 2005), 141–258.

11. It is worth noting that twelve of the senior wranglers at Cambridge between 1781 and 1807 were privately taught in the Sedbergh area by the self-taught mathematician and surgeon John Dawson.

12. Sheehan, *Enlightenment Bible*, 244–47; and Henry Gunning, *Reminiscences of the University, Town, and County of Cambridge: From the Year 1780*, 2nd ed. (London, 1855), 2:198–99.

13. Duncan Forbes, *The Liberal Anglican Idea of History* (Cambridge: Cambridge University Press, [1952] 2006), 29–30, 36–39; J. P. Powell, "Some Nineteenth-Century Views of the University Curriculum," *History of Education Quarterly* 5, no. 2 (June 1965); Perry Williams, "Passing on the Torch: Whewell's Philosophy and Principles of English Education," in *William Whewell: A Composite Portrait*, ed. Menachem Fisch and Simon Schaffer (Oxford: Oxford University Press, 1991), 120; and Simon Schaffer, "The History and Geography of the Intellectual World: Whewell's Politics of Language," in Fisch and Schaffer, *William Whewell*, 202. Whether this period was still an old regime or ushering in a new one is discussed in Boyd Hilton, *A Mad, Bad & Dangerous People? England 1783–1846* (Oxford: Oxford University Press, 2008), 24–31.

14. Whewell to James Garth Marshall, December 27, 1842, in Janet Mary Douglas, *The Life and Selections from the Correspondence of William Whewell* (London, 1881), 282; Richard Yeo, "Whewell, William (1794–1866)," *Oxford Dictionary of National Biography* (hereafter *ODNB*), https://www.oxforddnb.com/; Robert Hole, *Pulpits, Politics and Public Order in England 1760–1832* (Cambridge: Cambridge University Press, 1989), 249.

15. For an early but succinct overview of the Cambridge (primarily Trinity) and Oxford (mainly Oriel) approaches to theology, see Charles Richard Sanders, *Coleridge and the Broad Church Movement* (Durham, NC: Duke University Press, 1942), 14–15. For later studies, see, for example, William Gibson, *The Church of England 1688–1832: Unity and Accord* (London: Routledge, 2001), 2; and Pietro Corsi, *Science and Religion: Baden Powell and the Anglican Debate, 1800–1860* (Cambridge: Cambridge University Press, 1988), 198–99.

16. Harro Maas, "'A Hard Battle to Fight': Natural Theology and the Dismal Science, 1820–50," *History of Political Economy* 40, no. 5 (2008): 144.

17. Paul Oslington, "Natural Theology, Theodicy, and Political Economy in Nineteenth-Century Britain: William Whewell's Struggle," *History of Political Economy* 49, no. 4 (December 2017): 583–85.

18. Alec R. Vidler, *The Church in an Age of Revolution* (1961; repr., St. Ives, UK: Penguin Books, 1990), 17–32; Owen Chadwick, *The Victorian Church: Part One 1829–1859*, 3rd ed. (London: SCM Press, [1966] 1971), 2–3; William Gibson, *Church, State and Society, 1760–1850* (Basingstoke, UK: Palgrave Macmillan, 1994), 51. For the working class, see the still unsurpassed E. P. Thompson, *The Making of the English Working Class* (London: Victor Gollancz, 1963). Works on the history of the French Revolution are vast. For a good English-language introduction and overview, see William Doyle, *The Oxford History of the French Revolution*, 3rd ed. (Oxford: Oxford University Press, [1989] 2018).

19. Boyd Hilton, *The Age of Atonement: The Influence of Evangelicalism on Social and Economic Thought 1785–1865* (Oxford: Oxford University Press, 1988).

20. Gibson, *Church, State and Society*, 19–25, 28–35, 52–53; Richard Brent, *Liberal Anglican Politics: Whiggery, Religion, and Reform 1830–1841* (Oxford: Oxford University Press, 1987), 4–5.

21. Adrian Desmond, *The Politics of Evolution: Morphology, Medicine, and Reform in Radical London* (Chicago: University of Chicago Press, 1989), 7, 265.

22. Henry Brougham, quoted in Desmond, *Politics of Evolution*, 25; Susan Faye Cannon, *Science in Culture: The Early Victorian Period* (New York: Dawson, 1978), 18; Williams, "Passing on the Torch," 120; William J. Ashworth, "The Calculating Eye: Baily, Herschel, Babbage and the Business of Astronomy," *British Journal for the History of Science* 27, no. 4 (1994); James A. Secord, *Visions of Science: Books and Readers at the Dawn of the Victorian Age* (Oxford: Oxford University Press, 2014), 7–14. The establishment of London University was quickly followed by the founding of an Anglican London rival, King's College.

23. Nigel Leask, *The Politics of Imagination in Coleridge's Critical Thought* (Basingstoke, UK: Macmillan Press, 1988), 184.

24. Connop Thirlwall to John Candler, April 14, 1810, and July 1815, in Thirlwall, *Letters Literary and Theological of Connop Thirlwall*, ed. J. J. Stewart Perowne and Louis Stokes (London, 1881), 5–6, 23–24.

25. McFarland, "Early Literary Career," 44. For the importance of Horne Tooke to the rational study of language, see especially Hans Aarsleff, *The Study of Language in England, 1780–1860* (Minneapolis: University of Minnesota Press, [1967] 1983), chap. 3. For the prevailing hostility toward Locke and the general attack on rationalism, see, for example, Forbes, *Liberal Anglican Idea of History*, 30–44, 107–8.

26. Hare to Whewell, May 13, 1822, William Whewell Archive, Trinity College (hereafter TC): Add.Ms.a.77/127.

27. William Keach, "Romanticism and Language," in *The Cambridge Companion to British Romanticism*, ed. Stuart Curran (Cambridge: Cambridge University Press, 1993), 105.

28. John Hedley Brooke, *Science and Religion: Some Historical Perspectives* (Cambridge: Cambridge University Press, 1991), 167–68; and David Attis, *Mathematics and the Making of Modern Ireland: Trinity College Dublin from Cromwell to the Celtic Tiger* (Boston: Docent Press, 2014), 80–86.

29. John Rogerson, *Old Testament Criticism in the Nineteenth Century: England and Germany* (London: SPCK, 1984), 150–53, 159, 166.

30. Aileen Fyfe, "The Reception of William Paley's *Natural Theology* in the University of Cambridge," *British Journal for the History of Science* 30, no. 3 (September 1997); Matthew D. Eddy and David Knight, "Introduction," in William Paley, *Natural Theology or Evidence of the Existence and Attributes of the Deity, collected from the Appearances of Nature* (1802, Oxford: Oxford University Press, 2006), xiv, xxii, xxix.

31. Hole, *Pulpits, Politics and Public Order*, 78–81, 203, 256–57.

32. Samuel Clarke, quoted in Mayr, *Authority, Liberty and Automatic Machinery*, 100.

33. Maas, "'A Hard Battle to Fight,'" 147–48, 155.

34. Rogerson, *Old Testament Criticism*, 163–64.

35. Sheehan, *Enlightenment Bible*, 251–52.

CHAPTER ONE: TRINITY IN THE EARLY-NINETEENTH CENTURY

Epigraphs: Julius Charles Hare, "The Children of Light: A Sermon," Cambridge, 1828, in Hare, *Sermons Preacht on Particular Occasions* (Cambridge, 1858), 12; William Whewell, *Astronomy and General Physics Considered with Reference to Natural Theology* (London, 1833), 8.

1. Sheldon Rothblatt, "The Student Sub-Culture and the Examination System in Early Nineteenth Century Oxbridge," in *Oxford and Cambridge from the*

Fourteenth to the Early-Nineteenth Century, ed. Lawrence Stone (Princeton, NJ: Princeton University Press, 1975), 285–87, 292; Powell, "Some Nineteenth-Century Views of the University Curriculum," 97–109.

2. P. M. S. Dawson, "Poetry in an Age of Revolution," in Curran, *Cambridge Companion to British Romanticism*, 52.

3. Sedgwick, quoted in John Willis Clark, *The Life and Letters of the Reverend Adam Sedgwick* (1890; repr., London: Forgotten Books, 2015), 1:71.

4. Charles Henry Cooper, *Annals of Cambridge* (Cambridge, 1842–1853), 4:521.

5. Thorp to Whewell, May 17, 1822, TC: Add.Ms.a.213/131.

6. Jonathan Smith and Christopher Stray, eds., *Cambridge in the 1830s: The Letters of Alexander Chisholm Gooden, 1831–1841* (Woodbridge, UK: Boydell & Brewer, 2003), 9.

7. James Gooden to Alexander Gooden, November 7, 1836, in Smith and Stray, *Cambridge in the 1830s*, 56.

8. For Whewell's experiences 1,200 feet below the surface with Airy, see their *Account of Experiments made at Dolcoath Mine, I Cornwall, in 1826 and 1828, for the purpose of determining the Density of the Earth* (Cambridge, 1828); Whewell to Hare, June 4, 1826, TC: Add.Ms.a.215/11; and Whewell to John Stevens Henslow, June 9, 1826, TC: Add.Ms.a.206/799. See also Isaac Todhunter, *William Whewell, D. D., Master of Trinity College, Cambridge: An Account of His Writings with Selections from His Literary and Scientific Correspondence* (London, 1876), 1:37–38.

9. For Whewell's nickname, see Robert O. Preyer, "The Romantic Tide Reaches Trinity: Notes on the Transmission and Diffusion of New Approaches to Traditional Studies at Cambridge, 1820–1840," *Annals of the New York Academy of Sciences* 360, no. 1 (April 1981): 44.

10. Connop Thirlwall to John Candler, May 1810, in Perowne and Stokes, *Letters of Connop Thirlwall*, 9.

11. Michael Kent Havens, "Coleridge on the Evolution of Language," *Studies in Romanticism* 20, no. 4 (Winter 1981): 182–83.

12. Hugh James Rose to Whewell, October 5, 1822, TC: Add.Ms.a.211/134.

13. M. H. Abrams, *The Mirror and the Lamp: Romantic Theory and the Critical Tradition* (Oxford: Oxford University Press, [1953] 1971), 101–6; Forbes, *Liberal Anglican Idea of History*, 36–37, 40–41, 58; Richards, *Romantic Conception of Life*, 12–14; Stephen Prickett, *Coleridge and Wordsworth: The Poetry of Growth* (Cambridge: Cambridge University Press, 1970), 88; John Worthen, *The Cambridge Introduction to Samuel Taylor Coleridge* (Cambridge: Cambridge University Press, 2010), 104.

14. Prickett, *Coleridge and Wordsworth*, 92–101; and Maurice Cranston, *The Romantic Movement* (Oxford: Oxford University Press, 1994), 16–17.

15. The most well-known critique of the mathematics taught at Cambridge came from the forthright Scottish mathematician John Playfair, who made a

scathing attack on both Oxford and Cambridge in an 1808 review of Laplace's *Mechanique Celeste* for the *Edinburgh Review*; see Amy Ackerberg-Hastings, "Analysis and Synthesis in John Playfair's 'Elements of Geometry,'" *British Journal for the History of Science* 35, no. 1 (2002): 67, 70.

16. John Connop Thirlwall Jr., *Connop Thirlwall: Historian and Theologian* (London: Society for the Promoting Christian Knowledge, 1936), 15. For the prevailing context, see John Belchem, "Henry Hunt and the Evolution of the Mass Platform," *English Historical Review* 93, no. 369 (October 1978); and Malcolm Chase, *The People's Farm: English Radical Agrarianism 1775–1840* (Oxford: Oxford University Press, 1988), 18–120.

17. Thirlwall, *Connop Thirlwall*, 16; Cooper, *Annals of Cambridge*, 4:516–17; Distad, *Guessing at Truth*, 29; and Laura J. Snyder, *The Philosophical Breakfast Club: Four Remarkable Friends Who Transformed Science and Changed the World* (New York: Broadway Books, 2010), 65–66.

18. Whewell to Rose, March 25, 1817, TC: R.2.99/1.

19. Wood was repaid for his loyal services by being appointed dean of Ely in 1820—a position he retained until he died and was replaced by the Trinity don "Gentleman" Peacock. Cambridge colleges, none more so than Trinity, efficiently controlled the county parishes and represented a dominant force in county and town. Three years later French was rewarded by being appointed master of Jesus.

20. Joan L. Richards, *Mathematical Visions: The Pursuit of Geometry in Victorian England* (London: Academic Press, 1988), 2–3. There was nothing unusual in Whewell's position toward geometry as part of a divine language; see, for example, April G. Shelford, "Thinking Geometrically in Pierre-Daniel Huet's 'Demonstratio Evangelica' (1679)," *Journal of the History of Ideas* 63, no. 4 (2002). For Kant and geometry, see, for example, Frederick C. Beiser, *German Idealism: The Struggle against Subjectivism 1781–1801* (Cambridge, MA: Harvard University Press, 2002), 57.

21. Newton, quoted in Niccolo Guicciardini, *Reading the Principia: The Debate of Newton's Mathematical Methods for Natural Philosophy from 1687–1736* (1999; repr., Cambridge: Cambridge University Press, 2003), 107.

22. Ashworth, "Memory, Efficiency, and Symbolic Analysis," and "Analytical Society (act. 1812–1813)," *ODNB*.

23. David A. Valone, "Hugh James Rose's Anglican Critique of Cambridge Science, Antirationalism, and Coleridgean Idealism in Late Georgian England," *Albion: A Quarterly Journal Concerned with British Studies* 33, no. 2 (Summer 2001): 219. For Whewell and Rose's translation, see Whewell to Herschel, March 6, 1817, in Todhunter, *William Whewell*, 2:16; and Whewell to Rose, March 25, 1817, and April 15, 1817, TC: R.2.99/1 and R.2.99/2–3. The actual translation of Lacroix's *Elementary Treatise on the Differential and Integral Calculus* was by Babbage, Herschel, and Peacock and appeared in 1816.

24. Whewell to his father, February 6, 1816, in Douglas, *Life of Whewell*, 22; and Todhunter, *William Whewell*, 1:6–7.

25. For Peacock's nickname, see Preyer, "Romantic Tide Reaches Trinity," 44. It seems Bromhead's ancestor was one of the founders of Caius; see Whewell to De Morgan, February 8, 1857, in Todhunter, *William Whewell*, 2:409–10.

26. Snyder, *Philosophical Breakfast Club*. This confusion seems to stem from Walter F. Cannon's earlier attempt to place these men within a distinct Cambridge network; see his "Scientists and Broad Churchmen: An Early Victorian Intellectual Network," *Journal of British Studies* 4, no. 1 (1964), although Cannon does point to the TC and the importance of German idealism rather than Baconian empiricism in his "History in Depth: The Early Victorian Period," *History of Science* 3, no. 1 (1964): 23, 32.

27. Ashworth, "Memory, Efficiency, and Symbolic Analysis," 629, 632. Joan L. Richards shows this clear demarcation between Whewell and Herschel in her *Mathematical Visions*, 24–25.

28. Étienne Bonnot de Condillac, *Logic, or the First Development of the Art of Thinking* (Paris, 1780), in *Philosophical Writings of Étienne Bonnot de Condillac*, trans. Franklin Philip with Harlan Lane (London: Lawrence Erlbaum, 1982), 414. The link between Condillac, Lagrange, Laplace, and Lacroix are best described in Maria Panteki, "Relationships between Algebra, Differential Equations, and Logic in England, 1800–1860" (PhD diss., University of Middlesex, 1991), 59–60; and Panteki, "French 'Logique' and British 'Logic': On the Origins of Augustus De Morgan's Early Logical Inquiries, 1805–1835," in *Handbook of the History of Logic*, vol. 4, *British Logic in the Nineteenth Century*, ed. Dov M. Gabbay and John Woods (Amsterdam: Elsevier, 2008), 387–90, 396–97.

29. Tim Milnes, "Coleridge's Logic," in Gabbay and Woods, *Handbook of the History of Logic*, 47, 56.

30. Harvey W. Becher, "William Whewell and Cambridge Mathematics," *Studies in History and Philosophy of Science* 11, no. 1 (January 1980): 7, 10.

31. Babbage and Herschel, "Plan of a New Society," Herschel Papers, St. John's College, Cambridge, and quoted in Ashworth, "Memory, Efficiency, and Symbolic Analysis," 634. For mathematics in general during this period, see Becher, "William Whewell and Cambridge Mathematics"; and Harvey W. Becher, "Radicals, Whigs, and Conservatives: The Middle and Lower Classes in the Analytical Revolution in Cambridge in the Age of Aristocracy," *British Journal for the History of Science* 28, no. 4 (December 1995); Bloor, "Hamilton and Peacock on the Essence of Algebra," in *Social History of Nineteenth-Century Mathematics*, eds. Herbert Mehrtens, Henk Bos and Ivo Schneider (Boston: DA Information Services, 1981); Helena Pycior, "Early Criticism of the Symbolical Algebra," *Historica Mathematica* 9, no. 4 (November 1982); Phillip Enros, "The Analytical Society (1812–1813): Precursor of the Renewal of Cambridge Mathematics," *Historia Mathematica* 10, no. 1 (February 1983);

Crosbie Smith and M. Norton Wise, *Energy and Empire: A Biographical Study of Lord Kelvin* (Cambridge: Cambridge University Press, 1989), 149–76; Ivor Grattan-Guiness, "Charles Babbage: Production by Numbers," *Annals of Science* 47, no. 1 (1990); Joan L. Richards, "Rigor and Clarity: Foundations of Mathematics in France and England, 1800–1840," *Science in Context* 4, no. 2 (1991); M. V. Wilkes, "Herschel, Peacock, Babbage, and the Development of the Cambridge Curriculum," *Notes and Records of the Royal Society* 44, no. 2 (1990); Panteki, "Relationships between Algebra"; Andrew Warwick, *Masters of Theory: Cambridge and the Rise of Mathematical Physics* (Chicago: University of Chicago Press, 2003), 66–84; Menachem Fisch, "The Problematic History of Nineteenth-Century British Algebra," *British Journal for the History of Science* 27, no. 3 (September 1994).

32. Charles Babbage, "On the Influence of Signs in Mathematical Reasoning," *Transactions of the Cambridge Philosophical Society* 2 (1827, but read on December 16, 1821), in *The Works of Charles Babbage*, ed. Martin Campbell-Kelly (London: Pickering & Chatto, 1989), 1:372.

33. Jephson was later accused and cleared of having sex with another man; see William Cobbett, *The Parson and the Boy; Or, The Fire-Shovel Hats in a Bustle: Being an Account of the Trial of the Reverend THOMAS JEPHSON, Doctor of Divinity, and Tutor at Cambridge, on a Charge of attempting to commit an Unnatural Crime on JAMES WELSH, a labourer* (Cambridge, 1823).

34. Ashworth, "Memory, Efficiency, and Symbolic Analysis," and "Calculating Eye"; Simon Schaffer, "Paper and Brass: The Lucasian Professorship 1820–39," in *From Newton to Hawking: A History of Cambridge University's Lucasian Professors of Mathematics*, ed. Kevin Knox and Richard Noakes (Cambridge: Cambridge University Press, 2003).

35. Babbage to Herschel, November 13, 1815; Peacock to Herschel, March 4, 1817, Herschel Correspondence, St. John's College; Whewell to Herschel, March 6, 1817, in Todhunter, *William Whewell*, 2:16.

36. Anon., "Review of *A Treatise on Plane and Spherical Trigonometry*. By John Bonnycastle," *Eclectic Review*, January 1808, 58–59. This was a favorable review of Bonnycastle's work since it remained applied mathematics and was clearly demonstrated.

37. Mary Milner, *The Life of Isaac Milner* (London, 1848), 13, and quoted in Hopkins, *Charles Simeon*, 21.

38. Hopkins, *Charles Simeon*, 16–21; and Cooper, *Annals of Cambridge*, 4:543.

39. Ashworth, "Memory, Efficiency, and Symbolic Analysis," 637.

40. This was why Whewell, for example, got so upset with his former pupil, Babington Macaulay, who gave a speech in Edinburgh in 1846, "In Praise of Superficial Knowledge," where he argued human knowledge *and intelligence (thinking and reasoning)* had advanced through history. He lost his Edinburgh

seat the following year. For Whewell's critical comments, see, for example, his review of Macaulay in *Fraser's Magazine*, August 1849, 171–75.

41. John Herschel, "Great Physical Theories," n.d., Herschel Papers, Harry Ransom Center, University of Texas at Austin (hereafter Herschel Papers), quoted in Ashworth, "Memory, Efficiency, and Symbolic Analysis," 638–39. This was an ongoing debate since the eighteenth century concerning the perceived corruption of virtue via growing commerce; see, for example, J. G. A. Pocock, *Virtue, Commerce, and History: Essays on Political Thought and History, Chiefly in the Eighteenth Century* (Cambridge: Cambridge University Press, 1985).

42. Babbage, "On the Influence of Signs," 376; John Herschel, "Scraps of Philosophy," n.d., Herschel Papers, and "Mathematics," *Encylopaedia Metropolitana* (1830); the latter is reprinted in *Aspects of the Life and Thought of John Herschel*, ed. Silvian S. Schweber (New York: Arno Press, 1981), 434–35. This aspect of Babbage's thinking is underlined in Michael A. Lewis, "Charles Babbage: Reclaiming an Operations Management Pioneer," *Journal of Operations Management* 25, no. 2 (March 2007): 252, 255–56.

43. Desmond MacHale, *The Life and Work of George Boole: A Prelude to the Digital Age* (1985; repr., Cork: Cork University Press, 2014), 19–20.

44. MacHale, *Life and Work of George Boole*, 266–67; and Doron Swade, *The Cogwheel Brain: Charles Babbage and the Quest to Build the First Computer* (London: Little, Brown, 2000), 169.

45. George Boole, *The Mathematical Analysis of Logic* (Cambridge, 1847), and *An Investigation of the Laws of Thought, on which are Founded the Mathematical Theories of Logic and Probabilities* (London, 1854).

46. MacHale, *Life and Work of George Boole*, 58–61, 88–92. He also gained support from William Thompson.

47. George Boole, quoted in MacHale, *Life and Work of George Boole*, 42. Bromhead also went on to patronize the Nottingham baker's mathematical son, George Green (54). Whewell may have come across Boole's work in the *Cambridge Mathematical Journal* where he published several articles in the 1840s, but as far as I know, Whewell never mentioned him. By contrast, he probably knew Green, who at the age of nearly forty went to Caius and finished fourth wrangler.

48. Boole, *Mathematical Analysis of Logic*, 6, 8–9. For an overview of his logic, see especially Dale Jacquette, "Boole's Logic," in Gabbay and Woods, *Handbook of the History of Logic*, 332–39.

49. Boole, *Investigation of the Laws of Thought*, 14–15, 334, 345.

50. For example, using his development of "quaternions," William Rowan Hamilton at Trinity College, Dublin, attempted to defend the algebraic system and argued it was imbued with a spiritual existence; see Attis, *Mathematics and the Making of Modern Ireland*, 20–21; and Pycior, "Early Criticism," 408.

51. Henry Chadwick, "Romanticism and Religion," in his *Tradition and Exploration: Collected Papers on Theology and the Church* (Norwich: Canterbury, 1994), 225. For Berkeley and geometry, see Attis, *Mathematics and the Making of Modern Ireland*, 91.

52. John Belchem, *"Orator" Hunt: Henry Hunt and English Working Class Radicalism* (London: Breviary Stuff Publications, 2012).

53. Whewell to his father, June 6, 1816 in Douglas, *Life of Whewell*, 23. Extensive details of the protest are given at "Ely and Littleport Riots of 1816," *Wikipedia*, https://en.wikipedia.org/wiki/Ely_and_Littleport_riots_of_1816. For the history of this riotous year in East Anglia, see A. J. Peacock, *Bread or Blood: A Study of the Agrarian Riots in East Anglia in 1816 with a Foreword by E. P. Thompson* (London: V. Gollancz, 1965).

54. Cooper, *Annals of Cambridge*, 4:514.

55. Sheepshanks to Whewell, March 1, 1819, TC: Add.Ms.a.213/61.

56. Hilton, *Mad, Bad & Dangerous People?*, 247.

57. Henry Hallam to Whewell, November 13, 1845, TC: Add.Ms.a.205/86. Whewell was a fan of Hallam's constitutional history. He described him earlier to Jones as "an old whig and at times a stern one; but then he is an old whig not a new whig; and I see scarcely any points on which we the constitutional conservatives are called upon to differ with him"; Whewell to Jones, October 6, 1843, TC: Add.Ms.c.51/230, and reprinted in Todhunter, *William Whewell*, 2:318.

58. John B. Morrell, "Theophobia Gallica: Natural Philosophy, Religion, and Politics in Edinburgh, 1789–1815," *Notes and Records of the Royal Society of London* 26, no. 1 (June 1971): 44–45. However, his successor, John Playfair, was a supporter of Continental developments in mathematics; see, for example, Ackerberg-Hastings, "Analysis and Synthesis."

59. William Hamilton, *On the Study of Mathematics*, 2nd ed. (London, 1854), 217, and quoted in Richard Olson, *British Natural Philosophy and British Physics: A Study in the Foundations of the Victorian Style* (Princeton, NJ: Princeton University Press, 1975), 21. For Hamilton's logic and attempt to quantify the predicate, see Ralph Jessop, "The Logic of Sir William Hamilton: Tunnelling through Sand to Place the Keystone in the Aristotelic Arch," in Gabbay and Woods, *Handbook of the History of Logic*.

60. Thomas Young, "An Essay on Cycloidal etc.," *British Magazine*, 1800, quoted in Phillip Enros, "The Analytical Society: Mathematics at Cambridge University in the Early Nineteenth Century" (PhD diss., University of Toronto, 1979), 72.

61. Thomas Young, quoted in Ackerberg-Hastings, "Analysis and Synthesis," 68.

62. Whewell considered Hamilton's "metaphysics . . . as worthless as he was subtle and learned"; see Whewell to Forbes, February 29, 1860, in Todhunter,

William Whewell, 2:419. This form of criticism was common in nineteenth-century Britain; see Pycior, "Early Criticism."

63. Hawkins to Whewell, January 31, 1846, TC: Add.Ms.a.206/25.

64. Hawkins to Whewell, January 11, 1849, TC: Add.Ms.a.206/37.

65. Schaffer, "Babbage's Intelligence."

66. Adrian Desmond and James Moore, *Darwin* (London: Penguin, 1991), 213, 261. For the importance of the clockwork and subsequently the machine metaphor, see especially Mayr, *Authority, Liberty and Automatic Machinery.*

67. See, for example, Paley, *Natural Theology*, 12–14.

68. David Hume, "Of Miracles" (1748), in *Essays, Literary, Moral and Political by David Hume* (London, 1875).

69. Charles Babbage, *The Ninth Bridgewater Treatise: A Fragment*, vol. 9 of *The Works of Charles Babbage* (London, 1837), 8.

70. Babbage, *Ninth Bridgewater Treatise*, 9.

71. Richard Yeo, *Defining Science: William Whewell, Natural Knowledge and Public Debate in Early Victorian Britain* (Cambridge: Cambridge University Press, 1993), 3; and Cranston, *Romantic Movement*, 27–28. Although Forbes does not mention Whewell, but does Hare and Thirlwall, he was a prime example of Forbes's analysis of those who drew on a Liberal Anglican idea of history, see his *Liberal Anglican Idea of History.*

72. D. M. Peacock, *A Comparative View of the Principles of the Fluxional and Differential Calculus Addressed to the University of Cambridge* (Cambridge, 1819), 85, and quoted in J. M. Dubbey, *The Mathematical Work of Charles Babbage* (Cambridge: Cambridge University Press, 1978), 46–47.

73. William Whewell, *Thoughts on the Study of Mathematics, as a Part of a Liberal Education,*

(London, 1835), 42. For further discussion, see Becher, "William Whewell and Cambridge Mathematics," 30–31, 40.

74. Brent, *Liberal Anglican Politics*, 6; Hole, *Pulpits, Politics and Public Order*, 254.

75. James F. White, *The Cambridge Movement: The Ecclesiologists and the Gothic Revival* (Cambridge: Cambridge University Press, 1962), 3, 7.

76. I am grateful to Boyd Hilton for explaining the politics of the pew system to me.

77. Hopkins, *Charles Simeon*, 13.

78. Gibson, *Church, State and Society*, 34, 60, 93, 95. Whewell refers to hearing Simeon preach in November 1823, and rather tamely told his sister on the day Simeon died that "all that I have heard of him has interested me very much"; see his letter to his aunt and sisters, November 2, 1823, and to his sister, November 14, 1836, in Douglas, *Life of Whewell*, 93, 183.

79. John Gascoigne, "Mathematics and Meritocracy: The Emergence of the Cambridge Mathematical Tripos," *Social Studies of Science* 14, no. 4 (November

1984), and Gascoigne, *Cambridge in the Age of the Enlightenment: Science, Religion and Politics from the Restoration to the French Revolution* (Cambridge: Cambridge University Press, 1989).

80. McFarland, "Early Literary Career," 54; and G. F. McFarland, "Julius Hare: Coleridge, DeQuincey, and German Literature," *Bulletin of the John Rylands Library* 47, no. 1 (1964): 165.

81. Douglas, *Life of Whewell*, 28–33.

82. G. M. Trevelyan, *Trinity College: An Historical Sketch* (1943; repr., Cambridge: Cambridge University Press, 1990), 89.

83. For Friedrich Schlegel, see especially Richards, *Romantic Conception of Life*, chap. 2.

84. Whewell to Rose, August 30, 1817, TC: R.2.99/7.

85. Whewell to his sister, October 8, 1818, in Douglas, *Life of Whewell*, 47.

86. Peter B. Nockles, "Rose, Hugh James (1795–1838)," *ODNB*; and Gibson, *Church, State and Society*, 57.

87. Whewell was ecstatic at the appointment since it would mean he would meet his brother, William Wordsworth, and he wrote to Hare, "We will be the most poetical and psychological college in the universe"; see his letter dated July 1, 1820, in Todhunter, *William Whewell*, 2:37.

88. Richard Sharp, "Wordsworth, Christopher (1774–1846)," *ODNB*.

89. Gibson, *Church, State and Society*, 59, 97; and Corsi, *Science and Religion*, 14.

90. Cooper, *Annals of Cambridge*, 4:529; E. A. Smith, *A Queen on Trial: The Affair of Queen Caroline* (Stroud, UK: The History Press, 2016).

91. Cooper, *Annals of Cambridge*, 4:543; and Trevelyan, *Trinity College*, 93.

92. Christopher Stray, "A Parochial Anomaly: The Classical Tripos 1822–1900," in Smith and Stray, *Teaching and Learning in Nineteenth-Century Cambridge*, 31–44, 35–36; and Stray, "Curriculum and Style in the Collegiate University: Classics in Nineteenth-Century Oxbridge," *Histories of Universities* 16, no. 2 (2000): 189.

93. Stray, "Parochial Anomaly," 36.

94. Whewell to Jones, October 18, 1826, in Todhunter, *William Whewell*, 2:75.

95. "Dr. Wordsworth's Dinner Guests at Trinity Lodge 1820–1841," TC: 0.12.52.

96. Hare to Whewell, October 12, 1826, TC: Add.Ms.a.206/157.

97. Arthur Michael Ramsey, *F. D. Maurice and the Conflicts of Modern Theology* (Eugene, OR: WIPF & STOCK, [1951] 2011), 11.

98. For the Cambridge Apostles, see, for example, Richard Deacon, *The Cambridge Apostles: A History of Cambridge University's Elite Intellectual Secret Society* (London: R. Royce, 1985); and Peter Allen, *The Cambridge Apostles: The Early Years* (Cambridge: Cambridge University Press, 2010).

99. Faye, *Science in Culture*, 49.

100. Bunsen, quoted in J. W. Burrow, "The Uses of Philology in Victorian England," in *Ideas and Institutions of Victorian Britain: Essays in Honour of George Kitson Clark*, ed. Robert Robson and Walter F. Cannon (London: Bell, 1967), 195; Preyer, "Breaking Out," 62.

101. Havens, "Coleridge," 163–64.

102. Thirlwall, *Connop Thirlwall*, 41–48.

103. Bunsen, quoted in Burrow, "Uses of Philology," 196; and Stephen Prickett, *Words and the Word: Language, Poetics and Biblical Interpretation* (Cambridge: Cambridge University Press, 1988), 59.

104. Milnes, "Coleridge's Logic," 44.

105. H. J. Jackson, "Etymology and Etymologic," *Journal of the History of Ideas* 44, no. 1 (1983): 81.

106. Thirlwall, *Connop Thirlwall*, 19; and Forbes, *Liberal Anglican Idea of History*, 2.

107. Thirlwall, *Connop Thirlwall*, 28–30; Preyer, "Breaking Out," 61; Frederick Gregory, "Theology and the Sciences," in *Romanticism and the Sciences*, ed. Andrew Cunningham and Nicholas Jardine (Cambridge: Cambridge University Press, 1990), 71–73.

108. Robert O. Preyer, "John Stuart Mill on the Utility of Classical Greece," *Browning Institute Studies* 10 (1982): 48.

109. Richards, *Romantic Conception of Life*, 58, 102–3.

110. Henry Chadwick, "Romanticism and Religion," in his *Tradition and Exploration: Collected Papers on Theology and the Church* (Norwich, UK: Canterbury Press, 1994), 218–19.

111. Richards, *Romantic Conception of Life*, 46; Hole, *Pulpits, Politics, and Public Order*, 180–82; Worthen, *Coleridge*, 107–15. For the phrase "all-crushing," see Ernst Behler, "Introduction" to his *Fichte, Jacobi, and Schelling: Philosophy of German Idealism* (New York: Continuum, 1987), vii.

112. Frederick Maurice, ed., *The Life of Frederick Denison Maurice: Chiefly Told in His Own Letters* (London, 1884), 1:176, and quoted in Robert Robson and Walter F. Cannon, "Trinity College in the Age of Peel," in Robson and Cannon, *Ideas and Institutions*, 327.

113. Charles Richard Sanders, "Coleridge, Maurice and the Church Universal," *Journal of Religion* 21, no. 1 (January 1941): 31–34.

114. Whewell to Hare, February 13, 1835, in Todhunter, *William Whewell*, 2:202. For Coleridge's and Wordsworth's radical period, see especially Nicholas Roe, *Wordsworth and Coleridge: The Radical Years* (Oxford: Oxford University Press, 1988).

115. This perceived erosion of higher forms of knowledge by the rise of science is clearly described in Abrams, *Mirror and the Lamp*, 298–335.

116. Forbes, *Liberal Anglican Idea of History*, 5–11.

117. Hare, "Children of Light," 5–7.

118. Julius and Augustus Hare and Frederick Maurice, quoted in Distad, *Guessing at Truth*, 142, the Hare brothers on 47 and Frederick Maurice on 73. For the impact of Hare's sermon, see Brent, *Liberal Anglican Politics*, 163–64.

119. Julius Hare and Augustus Hare, *Guesses at Truth* (London, 1827), 121–23.

120. Stray, "Curriculum and Style," 190.

121. Distad, *Guessing at Truth*, 31–32, 39–43, 45–46; Cannon, "Scientists and Broad Churchmen," 77–78; Philip C. Rule, "Coleridge's Reputation as a Religious Thinker: 1816–1972," *Harvard Theological Review* 67, no. 3 (July 1974): 290. In 1824 Thirlwall, assisted by his friend Hare, translated two tales by the German poet and novelist Ludwig Tieck, and later that year pioneered the introduction of German theology into Britain by translating a critical essay on the Gospel of St. Luke by the theologian and philosopher Friedrich Schleiermacher. Kemble is given a prominent place in importing the new philology into England by Aarsleff, *Study of Language in England*, chap. 5.

122. Charles Merivale, *Autobiography and Letters of Charles Merivale, Dean of Ely*, ed. J. A. Merivale (Oxford, 1898), 98, and quoted in Thirlwall, *Connop Thirlwall*, 62; and Robson and Cannon, "Trinity College," 326.

123. R. C. Trench, *Letters and Memorials* (London, 1888), 1:10, and quoted in Robson and Cannon, "Trinity College," 26.

124. Thirlwall, *Connop Thirlwall*, 58.

125. Steffan Collini, *Matthew Arnold: A Critical Portrait* (Oxford: Oxford University Press, [1988] 1994), 105–7; Chadwick, "Romanticism and Religion," 224; Forbes, *Liberal Anglican Idea of History*, 13.

126. Jones to Whewell, August 13, 1822, TC: Add.Ms.c.52/6, quoted in Hilton, *Age of Atonement*, xx.

127. For Kant's transcendental metaphysics and religion, see, for example, Damon Linker, "From Kant to Schelling: Counter-Enlightenment in the Name of Reason," *Review of Metaphysics* 54, no. 2 (December 2000): 345–55; Terry Pinkard, *German Philosophy 1760–1860: The Legacy of Idealism* (Cambridge: Cambridge University Press, 2002), 19–44, 58–65; Gary Banham, *Kant's Transcendental Imagination* (Basingstoke, UK: Palgrave Macmillan, 2005). For the Coleridge quote, see Worthen, *Coleridge*, 12.

128. Whewell to Jones, August 16, 1822, TC: Add.Ms.c.51/14; and Whewell to Rose, September 24, 1822, TC: R.2.99/14. The letter to Jones is reprinted in Todhunter, *William Whewell*, 2:48–50.

129. Whewell to Jones, October 17, 1825, TC: Add.Ms.c.51/23.

130. Pinkard, *German Philosophy*, 85–96.

131. Pinkard, *German Philosophy*, 132–33.

132. Whewell acknowledges his debt to Schelling (and Kant) in his *Philosophy of the Inductive Sciences*, 1:x. For Schelling, see Pinkard, *German Philosophy*, 317–32. Eighteen forty-one was the same year the British government and the Prussian king authorized a Protestant bishop in Jerusalem and told Constanti-

nople to recognize Protestants in Turkey. This was a plan devised by Bunsen, and the bishop selected was Michael Solomon Alexander (1799–1845), a Prussian Jew and professor of Hebrew at King's College; see Chadwick, *Victorian Church*, 189–90.

133. Pinkard, *German Philosophy*, 330.

134. Linker, "From Kant to Schelling," 366.

135. Whewell to Jones, February 4, 1830, TC: Add.Ms.c.52/80.

136. Whewell to Jones, November 18, 1831, TC: Add.Ms.c.51/119.

137. William Whewell, *Astronomy and General Physics Considered with Reference to Natural Theology* (London, 1833), 328. He made the same argument in his *Philosophy of the Inductive Sciences*, 41–45.

138. Whewell, *Astronomy*, 3.

139. Whewell, *Astronomy*, 329, 334–39.

140. Hare to Whewell, June 30, 1827, TC: Add.Ms.a.206/170.

141. Whewell, *Astronomy*, 323–42.

142. Whewell, *Astronomy*, 329–30.

143. William Whewell, "Letter to Charles Babbage," *Athenaeum*, May 30, 1837, 1–7, 3–4. This was straight out of his Bridgewater Treatise; see, for example, 332.

144. Whewell, *Astronomy*, 8.

145. Charles Babbage, *The Ninth Bridgewater Treatise*, 2nd ed. (London, [1837] 1838), vii.

146. J. R. Garrard, "Thorp, Thomas (1797–1877)," *ODNB*.

147. Chadwick, *Victorian Church*, 213.

148. Hoxie N. Fairchild, "Romanticism and the Religious Revival in England," *Journal of the History of Ideas* 2, no. 3 (June 1941): 331–32, 335.

149. Alexandrina Buchanan, *Robert Willis and the Foundation of Architectural History* (Cambridge: Cambridge University Press, 2013), 74–75.

150. White, *Cambridge Movement*, 26–27, 50, 57.

151. Michael Hall, "What Do Victorian Churches Mean? Symbolism and Sacramentalism in Anglican Church Architecture, 1850–1870," *Journal of the Society of Architectural Historians* 59, no. 1 (2000): 80.

152. R. J. Smith, *The Gothic Bequest: Medieval Institutions in British Thought, 1688–1863* (Cambridge: Cambridge University Press, 1987), 156–57. For Gothic architecture and Whewell's philosophy, see, for example, Carla Yanni, "On Nature and Nomenclature: William Whewell and the Production of Architectural Knowledge in Early Victorian Britain," *Architectural History* 40 (1997); and Aleta Quinn, "William Whewell's Philosophy of Architecture and the Historicization of Biology," *Studies in History and Philosophy of Biological and Biomedical Science* 59 (October 2016): 13–15.

153. Augustus J. C. Hare, *Memorials of a Quiet Life* (London, 1872), 1:197, and quoted in Robson and Cannon, "Trinity College," 327.

154. Adrian Desmond and James Moore, *Darwin's Sacred Cause: Race, Slavery and the Quest for Human Origins* (London: Allen Lane, 2009), 31. This had implications for the intellectual development of other peoples around the world. For example, phrenologists started to claim parts of the brain in different peoples were different. South Americans, for example, had an enlarged faculty of Pride; Africans had significantly smaller faculties for Truth and Justice. In short, the shape of the head could reveal intelligence and the level of civilization (37).

155. Chadwick, "Romanticism and Religion," 219; and Chadwick, *Victorian Church*, 168.

156. Hilton, *Mad, Bad & Dangerous People?*, 468–75.

157. Coleridge, quoted in Worthen, *Coleridge*, 3.

158. Josef L. Altholz, *Anatomy of a Controversy: The Debate over "Essays & Reviews"* (Aldershot, UK: Scolar Press, 1994), 5.

159. Margaret Pawley, "Digby, Kenelm Henry (1795/6–1880)"; Sidney Lee, rev. Penelope Hicks, "Walker, (William) Sidney (1795–1846)"; J. R. Garrard, "Thorp, Thomas (1797–1877)"; Richard Smail, "Monk, James Henry (1784–1856)," all in *ODNB*.

160. This was a point acknowledged by Wordsworth, who wrote in his diary on being appointed master, "My reception from the Fellows was (especially I ought to mention Professor Monk) very kind"; "Diary of Christopher Wordsworth, 1819–1841," TC: 0.12.55. Most of the diary is concerned with Wordsworth's ailments, in particular, stomach disorders, bowel problems, and piles.

161. Robson and Cannon, "Trinity College," 318.

162. Whewell to Monk, June 2, 1823, in Todhunter, *William Whewell*, 2:51–53.

163. Whewell to Monk, 1841, and quoted in Todhunter, *William Whewell*, 1:xxi.

164. McFarland, "Early Literary Career," 55, 74.

165. Thirlwall, *Connop Thirlwall*, 54, 158–61.

166. Clark, *Life and Letters of the Reverend Adam Sedgwick*, 1:82.

167. G. Martin Murphy, "Malkin, Benjamin Heath (1769–1842)," *ODNB*.

168. Anon., "A Short Memoir of the Rev. Robert Edward Garnham," *Monthly Repository of Theology and General Literature*, vol. 10 (London, 1815). I am grateful to Jonathan Smith for this reference.

169. Chadwick, "Romanticism and Religion," 221–22; and Brooke, *Science and Religion*, 128–34.

170. Julius Charles Hare, *Charges to the Clergy of the Archdeaconry of Lewes, delivered at the Ordinary Visitations in the Years 1843, 1845 and 1846 with an anonymous Introduction* (Cambridge, 1856), xxiii.

171. Hare, "Children of the Light," 14.

CHAPTER TWO: PUTTING THE SOUL BACK IN THE MACHINE

Epigraph: Whewell to his sister, May 8, 1835, in Douglas, *Life of Whewell*, 173.

1. Charles Babbage, *Passages from the Life of a Philosopher*, vol. 11 of *Works of Charles Babbage* (1864; repr., London: Pickering and Chatto, 1991), 28-29.

2. Herbert Marsh, *An Inquiry and an Address to the Members of the Senate of the University of Cambridge, Occasioned by the Proposal to Introduce in This Place an Auxiliary Bible Society* (Cambridge, 1811), 10. For a closer look at the Cambridge auxiliary to the Bible Society during this period, see Kevin Knox, "Dephologisticating the Bible: Natural Philosophy and Religious Controversy in Late Georgian Cambridge," *History of Science* 34 (1996); Gunning, *Reminiscences*, 259–60; and Hole, *Pulpits, Politics and Public Order*, 191.

3. Gibson, *Church, State and Society*, 56.

4. Clark, *Life and Letters of the Reverend Adam Sedgwick*, 1:3.

5. Hole, *Pulpits, Politics and Public Order*, 190–93; Hilton, *Age of Atonement*, 3; Desmond and Moore, *Darwin's Sacred Cause*, 8.

6. Herbert Marsh, *Consequences of Neglecting to Give the Prayer Book with the Bible* (Cambridge, 1812), 55; and Isaac Milner, *Strictures on Some of the Publications of the rev. Herbert Marsh, Intended as a Reply to His Objections against the British and Foreign Bible Society* (London, 1813), 381. For Isaac Milner, see Kevin C. Knox, "The Negative Side of Nothing: Edward Waring, Isaac Milner and Newtonian Values," in Knox and Noakes, *From Newton to Hawking*. Milner's brother was Reverend Joseph Milner who, between 1794 and 1797, wrote *The History of the Church of Christ* in three volumes. He was also headmaster of Hull Grammar School and taught George Pryme (Trinity) and Thomas Perronet Thompson (Queen's).

7. Cooper, *Annals of Cambridge*, 4:500.

8. Gunning, *Reminiscences*, 31, 100; and Robson and Cannon, "Trinity College," 313.

9. Robert K. Forrest, "Marsh, Herbert (1757–1839)," *ODNB*; Sharp, "Wordsworth"; W. H. G. Armytage, *Four Hundred Years of English Education*, 2nd ed. (Cambridge: Cambridge University Press, [1964] 1970), 90–91.

10. For a clear overview of the central beliefs, and differences, of evangelicals during this period, see Hilton, *Age of Atonement*, 8–35.

11. Babbage, *Passages from the Life of a Philosopher*, 20. See also Bloor, "Hamilton and Peacock," 202–31.

12. Thomas Paine, *The Age of Reason*, pt. 1 (1794), in *The Selected Work of Tom Paine*, ed. Howard Fast (London: Bodley Head, 1948), 300.

13. The importance and long history of the Fall to notions of knowledge are best told in Harrison, *Fall of Man*.

14. Margaret Herschel to Dr. J. Stuart, May 4, 1837, in *Lady Herschel: Letters from the Cape, 1834–1838*, ed. Brian Warner (Cape Town: Friends of the South African Library, 1991), 132–33. For Herschel's theological doubts and republican tendencies at Cambridge, see, for example, S. S. Schweber, "John Herschel and Charles Darwin: A Study in Parallel Lives," *Journal of the History of Biology* 22, no. 1 (Spring 1989): 11–13.

15. For a succinct, critical overview of such an intellectual assault on this perspective, see, for example, Whewell, *On the Philosophy of Discovery*, 201–37.

16. William Whewell, "Diary: 23 May 1817 to May 1830" (hereafter Whewell diary), November 13, 1825, TC: R.18.9[13].

17. Whewell diary, November 13, 1825, TC: R.18.19[13].

18. William Whewell, "Sermons," probably given in Trinity Chapel, 1825, TC: R.6.17[22].

19. This analogy draws upon Hare, "Children of the Light"; and later Abrams, *Mirror and the Lamp*.

20. Cranston, *Romantic Movement*, 50–52; David Simpson, "Romanticism, Criticism and Theory," in Curran, *Cambridge Companion to British Romanticism*, 3; Dawson, "Poetry in an Age of Revolution," 57, 65; Smith, *Gothic Bequest*, 113–25.

21. Smith, *Gothic Bequest*, 153; Hole, *Pulpits, Politics and Public Order*, 253; Prickett, *Coleridge and Wordsworth*, 15.

22. Simon Schaffer, "States of Mind: Enlightenment and Natural Philosophy," in *The Languages of Psyche: Mind and Body in Enlightenment Thought*, ed. G. S. Rousseau (Berkley: University of California Press, 1987); and Desmond, *Politics of Evolution*, 107–22.

23. Ashworth, "Memory, Efficiency and Symbolic Analysis." During the 1820s the Savilian Professor of Geometry at Oxford University, Baden Powell, held Cambridge up to be the antithesis of the moribund and scientific poverty at his institution; see Corsi, *Science and Religion*, 36–41.

24. William Whewell, *An Elementary Treatise on Mechanics* (Cambridge, 1819), vi.

25. Herschel to Whewell, December 1, 1819, TC: Add.Ms.a.207/5.

26. Ashworth, "Calculating Eye."

27. Whewell, *Thoughts on the Study of Mathematics*, 3, 9, 12–13, 34.

28. Whewell to Mr. Wilkinson, Keswick, September 5, 1821, and to his aunt and sisters, Kendal, August 25, 1824, in Douglas, *Life of Whewell*, 66–67, 96–97.

29. Peter Thorsley, "German Romantic Idealism," in Curran, *Cambridge Companion to British Romanticism*, 76. Whewell, quoted in Laura J. Snyder, *Reforming Philosophy: A Victorian Debate on Science and Society* (Chicago: University of Chicago Press, 2006), 40. Snyder succinctly explains the role of Whewell's notion of ideas in "'The Whole Box of Tools': William Whewell and the Logic of Induction," in Gabbay and Woods, *Handbook of the History of Logic*, 179–81.

30. Thirlwall, *Connop Thirlwall*, 26–27.

31. Whewell to his sister Anne, January 1, 1822, in Douglas, *Life of Whewell*, 72–73.

32. Prickett, *Coleridge and Wordsworth*, 2, 27, 29, 37, 40, 44, 75–77; Beiser, *German Idealism*, 1–60.

33. This re-posturing between traditional religion and the rise of political economy is examined in nineteenth-century literature by Supritha Rajan in her *A Tale of Two Capitalisms: Sacred Economics in Nineteenth-Century Britain* (Ann Arbor: University of Michigan Press, 2015).

34. Anon., "Introduction," Julius Charles Hare, *Charges to the Clergy of the Archdeaconry of Lewes, Delivered at the Ordinary Visitations in the Years 1843, 1845, 1846* (Cambridge, 1856), xvi, xxi, xxv, xxx. Babbage, quoted in Swade, *Cogwheel Brain*, 9.

35. William Whewell, "Lecture 1: Definition of a System of Morality," in his *Lectures on Systematic Morality* (London, 1846; repr., Memphis, TN: General Books, 2012), 1–8, 2.

36. Rule, "Coleridge's Reputation," 308, 318; Cranston, *Romantic Movement*, 36; Worthen, *Coleridge*, 20.

37. Coleridge, quoted in Abrams, *Mirror and the Lamp*, 118.

38. Whewell diary, June 15 and 16, 1817, TC: $R.18.9^{2}$.

39. Whewell diary, November 26, 1820, TC: $R.18.9^{8}$.

40. Abrams, *Mirror and the Lamp*, 119, 124, 172.

41. Coleridge, quoted in Abrams, *Mirror and the Lamp*, 177; and Prickett, *Coleridge and Wordsworth*, 115.

42. Abrams, *Mirror and the Lamp*, 218–25; and Cranston, *Romantic Movement*, 21–47. Coleridge's emphasis on Shakespeare has led some to accuse him of plagiarizing the work of August Wilhelm von Schlegel, brother of Friedrich Schlegel; see Worthen, *Coleridge*, 90; and Prickett, *Coleridge and Wordsworth*, 63. August Schlegel was the first professor of Sanskrit in Europe. Forbes clearly demonstrated how popular this approach to history was among certain Anglican practitioners, including Hare and Thirlwall, in his *Liberal Anglican Idea of History*.

43. For a clear condemnation by Whewell of the purely deductive approach, see, for example, his *History of the Inductive Sciences*, 1:viii, 12.

44. Jeremy Bentham, *Church-of-Englandism and Its Catechism Examined* (London, 1818).

45. Valone, "Hugh James Rose's Anglican Critique," 225–27.

46. Douglas, *Life of Whewell*, 86–87.

47. Whewell to Lady Malcolm, October 28, 1827, in Douglas, *Life of Whewell*, 115.

48. Douglas, *Life of Whewell*, 121, 128.

49. Cooper, *Annals of Cambridge*, 4:553, 561–62.

50. Henry Gunning, *Reminiscences*, 2:368.

51. Thirlwall, "Works which Remain, and Works which Follow Us," a sermon preached in Charterhouse School on Founders Day, December 12, 1845, in *Remains Literary and Theological of Connop Thirlwall*, ed. J. J. Stewart Perowne (London, 1878), 3:315–16.

52. Julius Charles Hare, "Sermon II: Faith, A Practical Principle," n.d., in his *Victory of Faith and Other Sermons* (Cambridge, 1840; repr., Memphis, TN: General Books, 2012), 8–9, 11–13.

53. For the rise of commerce and its impact on literature during this period, see, for example, Rajan, *Tale of Two Capitalisms*.

54. Maria Hare, quoted in Distad, *Guessing at Truth*, 117.

55. Julius Charles Hare, "Sermon III: Office and Province of Faith," n.d., in his *Victory of Faith*, 17, 19, 23.

56. William Whewell, "Lecture VI: Virtue, Happiness, Conscience and the Will of God," in his *Lectures on Systematic Morality*, 38–54.

CHAPTER THREE: GERMAN BIBLICAL CRITICISM AND GUESSING

Epigraph: William Whewell, "Sermons," probably given in Trinity Chapel, February 18, 1827, TC: R.16.17[25].

1. Johann David Michaelis, *Introduction to the New Testament, Tr., and Augmented With Notes (And a Dissertation On the Origin and Composition of the Three First Gospels) by Herbert Marsh*, 4 vols. (Cambridge, 1801).

2. Distad, *Guessing at Truth*, 86–87; and Thirlwall, *Connop Thirlwall*, 29, 188.

3. Pinkard, *German Philosophy*, 149.

4. *Frederich Schleiermacher, A Critical Essay on the Gospel of St. Luke. By Dr. Frederich Schleiermacher. With an Introduction by the Translator* [Connop Thirlwall]. *Containing an Account of the Controversy respecting the Origin of the Three First Gospels since Bishop Marsh's Dissertation* (London, 1825), xi–xii. For Schleiermacher's "true church," see Pinkard, *German Philosophy*, 152–58.

5. Leask, *Politics of Imagination*, 106.

6. Rose, quoted in Rogerson, *Old Testament Criticism*, 166. For an old, but useful, overview of the Oxford Tracts and Tractarians, see Dean Church, *The Oxford Movement 1833–1845* (London, 1897), 92–145; and Hilton, *Mad, Bad & Dangerous People?*, 468–75. For Rose and his relationship to the Tractarians, see Geoffrey Faber, *Oxford Apostles: A Character Study of the Oxford Movement* (1933; repr., London: Faber and Faber, 1974), 263–64, 336–37, 348–49.

7. Rogerson, *Old Testament Criticism*, 9–10.

8. Thirlwall to Hare, October 28, 1823, and to John Thirlwall, July 23, 1829, in Thirlwall, *Letters of Thirlwall*, 70, 95. For the work of W. M. L. de Wette, see Rogerson, *Old Testament Criticism*, 28–49. The centrality of the German critical approach to Liberal Anglican history during this period is emphasized by Forbes in his *Liberal Anglican Idea of History*.

9. Rose, quoted in Rogerson, *Old Testament Criticism*, 166.

10. Hengstenberg, quoted in Rogerson, *Old Testament Criticism*, 86.

11. Hugh James Rose, *The State of the Protestant Religion in Germany: In a Series of Discourses Preached before the University of Cambridge* (Cambridge, 1825), xiii–xiv, 1, 3–7.

12. Rogerson, *Old Testament Criticism*, 27; and Preyer, "John Stuart Mill," 49. Rose's stance is historically situated in Sheehan, *Enlightenment Bible*, 248–49.

13. Rose, *State of the Protestant Religion in Germany*, 11, 14, 20–24, 27–28, 67–69, 102–3. The reason those spearheading the early Oxford Movement were also known as Tractarians was due to their publishing ninety theological tracts (*Tracts for the Times*) from 1833 to 1841. Owen Chadwick describes these tracts as "sharp-shooters" rather than "regular troops" in his *Victorian Church*, 74.

14. McFarland, "Early Literary Career," 74. For Simeon and Scholefield, see Hopkins, *Charles Simeon*, 97, 115.

15. Whewell to Rose, December 29, 1823, TC: R.2.99/21.

16. Whewell to Rose, November 19, 1826, TC: R.2.99/26.

17. Whewell to Rose, December 12, 1826, TC: R.2.99/27, and reprinted in Todhunter, *William Whewell*, 2:77–79.

18. Whewell to James Garth Marshall, December 27, 1842, in Douglas, *Life of Whewell*, 281–82.

19. Whewell to Rose, December 9, 1827, TC: R.2.99/31.

20. Whewell, "Sermons," probably given in Trinity Chapel, February 4, 1827, TC: R.6.17[25].

21. Whewell, "Sermons," February 4, 1827.

22. Whewell, "Sermons," probably given in Trinity Chapel, February 18, 1827.

23. See Harrison, *Fall of Man*.

24. Julius Hare and Augustus Hare, quoted in Distad, *Guessing at Truth*, 94–99; with the argument underlined in Forbes, *Liberal Anglican Idea of History*.

25. Brent, *Liberal Anglican Politics*, 180–83; Geoffrey N. Cantor, "Between Rationalism and Romanticism: Whewell's Historiography of the Inductive Sciences," in Fisch and Schaffer, *William Whewell*, 75–76; David B. Wilson, "William Whewell, Galileo, and Reconceptualizing the History of Science and Religion," *Notes and Records of the Royal Society* 65, no. 4 (December 2011): 345–46.

26. Renate Bridenthal, "Was There a Roman Homer? Niebuhr's Thesis and Its Critics," *History and Theory* 11, no. 2 (1972); Solomon Katz, "Remembrance of Things Past," *Pacific Historical Review* 38, no. 1 (February 1969): 10.

27. McFarland, "Julius Hare," 175.

28. Whewell to Jones, August 5, 1834, TC: Add.Ms.c.51/174. See also Whewell's later discussion on language and science in his *Philosophy of the Inductive Sciences*, 1:xlviii–cxx.

29. Whewell to Jones, August 21, 1834, TC: Add.Ms.c.51/175, and reprinted in Todhunter, *William Whewell*, 2:185–88.

30. Mill, quoted in Snyder, *Reforming Philosophy*, 204–5, 215.

31. This is an excellent example of Albert O. Hirschman's "perversity" thesis in his *Rhetoric of Reaction: Perversity, Futility and Jeopardy* (Cambridge, MA: Belknap Press, 1991), 12–19. This argument is perhaps most fully developed in Theodor W. Adorno and Max Horkheimer, *Dialectic of Enlightenment* (London: Verso, [1944] 2008).

32. Thirlwall to John Candler, January 17, 1817, and Thirlwall to Chevalier Bunsen, January 20, 1823, both in Thirlwall, *Letters of Thirlwall*, 40, 65–66.

33. Cymbre Quincy Raub, "Robert Chambers and William Whewell: A Nineteenth-Century Debate over the Origin of Language," *Journal of the History Ideas* 49, no. 2 (June 1988); and especially James A. Secord, *Victorian Sensation: The Extraordinary Publication, Reception, and Secret Authorship of Vestiges of the Natural History of Creation* (Chicago: University of Chicago Press, 2000), 226–31.

34. Whewell to Myers, March 16, 1845, in Douglas, *Life of Whewell*, 318. Herschel was not impressed by Whewell's book; see Whewell to Herschel, March 12, 1845, in Todhunter, *William Whewell*, 2:325.

35. Whewell, quoted in Raub, "Chambers and Whewell," 297. For Whewell's hostile critique of Locke and French sensory empiricists, see, for example, his *On the Philosophy of Discovery*, 201–12.

36. Aarsleff, *Study of Language*, 224–25.

37. Whewell to Jones, July 18, 1845, TC: Add.Ms.c.51/241. The history and reception of *Vestiges* is best told in Secord, *Victorian Sensation*.

38. Peter P. Liddel, "Introduction," in Connop Thirlwall, *History of Greece*, ed. Peter P. Liddel (1835; repr., Liverpool: Bristol Phoenix Press, 2007), ix–xxxii, xvi; Thirlwall, *Connop Thirlwall*, 41; Forbes, *Liberal Anglican Idea of History*, 12–60; Preyer, "Romantic Tide Reaches Trinity," 48.

39. Distad, *Guessing at Truth*, 108. By this stage, German Romanticism had engulfed German universities; see Cranston, *Romantic Movement*, 46.

40. Julius Charles Hare, *A Vindication of Niebuhr's History of Rome from the Charges of the Quarterly Review* (Cambridge, 1829), 9–11.

41. Hare, *Vindication of Niebuhr's History of Rome*, 15.

42. Distad, *Guessing at Truth*, 112.

43. Hare to Whewell, June 21, 1844, TC: Add.Ms.a.77/138.

44. Distad, *Guessing at Truth*, 111; Forbes, *Liberal Anglican Idea of History*, 16–19; Thirlwall, *Connop Thirlwall*, 96–100. Thirlwall's history was translated into German by L. Haymann and Leonhard Schmitz in Bonn; see Thirlwall, *Letters of Thirlwall*, 152.

45. Hare, *Vindication of Niebuhr's History of Rome*, 21.

46. Bridenthal, "Was There a Roman Homer?," 195–97.

47. Distad, *Guessing at Truth*, 111–12.

48. Hare, *Vindication of Niebuhr's History of Rome*, 26.

49. Hare, *Vindication of Niebuhr's History of Rome*, 30.

50. Julius Charles Hare, "Sermon IV: Power of Faith in Man's Natural Life," in Hare, *Victory of Faith*, 25–36, 35.

51. Niebuhr, quoted in Hare, *Vindication of Niebuhr's History of Rome*, 27.

52. Worthen, *Coleridge*, 84–89.

53. Thirlwall to Bunsen, November 21, 1831, December 16, 1832, and May 5, 1832, all in Thirlwall, *Letters of Thirlwall*, 101–2, 103–4, 107–8. For the widespread sinews of Christian evangelical belief, see especially Hilton, *Age of Atonement*.

54. Liddel, "Introduction," xviii; Thirlwall, *Connop Thirlwall*, 63; Distad, *Guessing at Truth*, 74; and McFarland, "Julius Hare," 184–85.

55. John Edwin Sandys, *A History of Classical Scholarship: The Eighteenth Century in Germany and the Nineteenth Century in Europe and the United States of America* (1903–1908; repr., Cambridge: Cambridge University Press, 2010), 3:400.

56. Christopher Stray, "One Museum to Another: The 'Museum Criticum' (1813–26) and the 'Philological Museum' (1831–33)," *Victorian Periodicals Review* 37, no. 3 (Fall 2004): 291, 293; N. Merrill Distad, "The Philological Museum of 1831–1833," *Victorian Periodicals Newsletter* 18 (December 1972).

57. Stray, "One Museum to Another," 290, 299–302.

58. Niebuhr, quoted in Preyer, "Romantic Tide Reaches Trinity," 52.

59. This approach was very typical of Coleridge's idea of history; see Michael John Kooy, "Romanticism and Coleridge's Idea of History," *Journal of the History of Ideas* 60, no. 4 (October 1999): 719–20, 722–25.

60. Burrow, "Uses of Philology," 183. This is the main point in Forbes, *Liberal Anglican Idea of History*.

61. Michael J. Franklin, *'Orientalist Jones': Sir William Jones, Poet, Lawyer, and Linguist, 1746–1794* (Oxford: Oxford University Press, 2011); Distad, *Guessing at Truth*, 79; Richards, *Romantic Conception of Life*, 21; Aarsleff, *Study of Language in England*, chap. 4.

62. Liddel, "Introduction," xxi.

63. Burrow, "Uses of Philology," 184–85, 197–202.

64. Jackson, "Etymology," 83.

65. Simon Schaffer, "Oriental Metrology and the Politics of Antiquity in Nineteenth-Century Survey Sciences," *Science in Context* 30, no. 2 (2017).

66. H. F. Augstein, "Prichard, James Cowles (1786–1848)," *ODNB*.

67. Desmond and Moore, *Darwin's Sacred Cause*, 54–56, 65; and Whewell to Lyell, June 5, 1863, in Todhunter, *William Whewell*, 2:431.

68. Farrar's first important publication was *An Essay on the Origin of Language* (1860), which drew on the work of the French philologist Ernest Renan

(1823–1892) and emphasized the phonetic resemblances of primitive words and their evolutionary development. This stimulated both Tennnyson and Darwin—the latter proposing Farrar as a fellow of the Royal Society in 1866. A decade later Benjamin Disraeli appointed him rector of St. Margaret's church and cannon of Westminster just next door. His sermons drew huge audiences; indeed, he was ranked the fifth greatest living English preacher in 1884. Finally, in 1895 he was made the dean of Canterbury. See Norman Vance, "Farrar, Frederic William (1831–1903)," *ODNB*.

69. Burrow, "Uses of Philology," 192; Richard Garnett, rev. Richard Smail, "Donaldson, John William (1811–1861)," ODNB.

70. Burrow, "Uses of Philology," 193.

71. Aarsleff, *Study of Language*, 211–14; and Preyer, "Breaking Out," 67.

72. Whewell to Hare, February 14, 1844, TC: Add.Ms.a.215/74.

73. Latham, quoted in Burrow, "Uses of Philology," 192.

74. William Whewell, "Mathematical Exposition of Some Doctrines of Political Economy," *Transactions of the Cambridge Philosophical Society* 3 (1830), and "Mathematical Exposition of Some Leading Doctrines in Mr Ricardo's Principles of Political Economy," *Transactions of the Cambridge Philosophical Society* 4 (1833). This aspect of Whewell's work is discussed in Snyder, *Reforming Philosophy*, 271, 281–85; and Oslington, "Natural Theology," 583–84.

75. Whewell to Jones, March 5, 1829, TC: Add.Ms.c.51/62.

76. Whewell to Jones, April 2, 1829, TC: Add.Ms.c.51/63.

77. Whewell to James Garth Marshall, December 27, 1842, in Douglas, *Life of Whewell*, 280.

78. A. W. Richeson, "On Faraday's Terminology in Electrolysis," *Isis* 36, no. 3–4 (1946); S. Ross, "Faraday Consults the Scholars: The Origins of the Terms of Electrochemistry," *Notes and Records of the Royal Society of London* 16, no. 2 (November 1961); R. Robson and Walter F. Cannon, "William Whewell, FRS (1794–1866)," *Notes & Records of the Royal Society* 19, no. 2 (December 1964): 181.

79. Robert O. Preyer, "The Language of Discovery: William Whewell and George Eliot," *Browning Institute Studies* 16 (1988): 128; and Jackson, "Etymology," 86.

80. Schaffer, "Whewell's Politics of Language," 208–9.

81. The anonymous pupil is quoted in Hare, *Charges to the Clergy of the Archdeaconry of Lewes*, vii–ix.

82. Desmond and Moore, *Darwin*, 70–73.

CHAPTER FOUR: TRINITY AND RELIGIOUS DISSENTERS

1. Chadwick, *Victorian Church*, 7–47. These events represent for J. C. D. Clark the real dawning of radical reform in Britain; see his *English Society 1688–1832* (Cambridge: Cambridge University Press, 1985). For a complimentary view of eighteenth-century radicalism as an essentially religious movement, see,

for example, Gibson, *Church of England*, 10–15. For a balanced interpretation, see Hilton, *Mad, Bad and Dangerous People?*, 420–38.

2. Jones to Whewell, October 11, 1832, TC: Add.Ms.c.51/142.

3. Whewell to Jones, October 11, 1832, TC: Add.Ms.c.51/142.

4. Thirlwall, *Connop Thirlwall*, 185.

5. Smith, *Gothic Bequest*, 175–78, 191–92; and Hole, *Pulpits, Politics and Public Order*, 124–26.

6. Gibson, *Church, State and Society*, 106–7.

7. Chadwick, *Victorian Church*, 46–47, 81–89, 142, 146–47; Brent, *Liberal Anglican Politics*, 6–10, 15, 190. Church rates were not abolished until 1868.

8. Captain Swing to William Whewell, n.d., but the paper is watermarked 1828, TC: R.1.76/45. I am grateful to Jonathan Smith for this reference.

9. Jones to Whewell, January 10, 1832, TC: Add.Ms.c.52/19.

10. Whewell to Jones, December 7, 1830, TC: Add.Ms.c.51/93.

11. Richard Jones, *Literary Remains, Consisting of Lectures and Tracts on Political Economy of the Late Rev. Richard Jones*, ed. John Cazenove, intro. William Whewell (London, 1859), 408–9.

12. Thirlwall to Bunsen, May 5, 1833, and October 10, 1833, in Thirlwall, *Letters of Thirlwall*, 107–8, 109–10.

13. For Pryme and his diffusion of Adam Smith's *Wealth of Nations* using Dugald Stewart's lectures as his model, see Shin Kubo, "George Pryme, Dugald Stewart, and Political Economy at Cambridge," *History of Political Economy* 45, no. 1 (2013).

14. Cooper, *Annals of Cambridge*, 4:581–83.

15. Duke of Gloucester, *Hansard*, April 21, 1834, 982.

16. Turton was also the dean of Peterborough (1830–1842), subsequently the dean of Westminster (1842–1845), and finally bishop of Ely (1845–1864).

17. Thirlwall, *Letters of Thirlwall*, 113–17.

18. Thirlwall to Bunsen, October 10, 1833, in Thirlwall, *Letters of Thirlwall*, 111–12.

19. Brent, *Liberal Anglican Politics*, 184–86. For the Oxford context, see Corsi, *Science and Religion*, 98.

20. Hugh James Rose, *The Duty of Maintaining the Truth* (Cambridge: Cambridge University Press, 1834), 15, 23.

21. Christopher Wordsworth to the resident fellows of Trinity, Cambridge, May 26, 1834, in Thirlwall, *Letters of Thirlwall*, 118–19. See also Liddel, "Introduction," ix–xxxii, xiv–xv; and Brent, *Liberal Anglican Politics*, 208–9.

22. J. W. Clark and H. C. G. Matthew, "Thirlwall, (Newell) Connop (1797–1875)," *ODNB*.

23. Trevelyan, *Trinity College*, 91, 93.

24. "Diary of Christopher Wordsworth, 1819–1841," February 3, 1821, TC: 0.12.55. Coddington was a newly appointed fellow and sub-tutor. He was senior

wrangler and first Smithsman in 1820, as well as being an astronomer, competent classicist, and linguist.

25. Robson and Cannon, "Trinity College," 330.

26. Hare to Whewell, June 1, 1834, TC: Add.Ms.a.206/162.

27. Hare to Whewell, July 4, 1834, TC: Add.Ms.a.215/32.

28. Whewell to Jones, March 24, 1834, TC: Add.Ms.c.51/164.

29. Thirlwall to George Pryme, June 2, 1834, in Thirlwall, *Letters of Thirlwall*, 121.

30. Perry was made deacon by Monk, the former Cambridge Professor of Greek and now bishop of Gloucester, in June 1833. He then went on to be the first bishop of Melbourne in Australia in 1847 and was the key player in establishing Trinity College at the University of Melbourne in 1860 to provide training for the clergy. In addition, he founded the Melbourne grammar school (1848) and Geelong grammar school (1857) to educate future young colonial leaders into a sound liberal and Anglican Christian education. He resigned in 1876 and came back to Britain to become, among other appointments, an active Church Missionary Society vice president and a committee member of both the Society for the Propagation of the Gospel and the Society for the Promotion of Christian Knowledge. See Robert S. M. Withycombe, "Perry, Charles (1807–1891)," *ODNB*.

31. Whewell to Sedgwick, May 27, 1834, in Douglas, *Life of Whewell*, 164.

32. William Whewell, *On the Admission of Dissenters to Academical Degrees* (Cambridge, 1834), 13, 18, and *Additional Remarks on Some Parts of Mr Thirlwall's Two Letters on the Admission of Dissenters to Academical Degrees* (Cambridge, 1834), 6–7, 15.

33. Rose, *Duty of Maintaining Truth*, 24–25.

34. Rose to Whewell, May 31, 1834, TC: Add.Ms.a.211/145.

35. Rose, quoted in Todhunter, *William Whewell*, 1:xxiii.

36. R. W. Church, *The Oxford Movement: Twelve Years 1833–1845* (London, 1897), 94–104.

37. Chadwick, *Victorian Church*, 69–75.

38. Chadwick, *Victorian Church*, 170–71, 179–81; Church, *Oxford Movement*, 201–17.

39. Whewell, quoted in Thirlwall, *Letters of Thirlwall*, 123.

40. Brent, *Liberal Anglican Politics*, 194.

41. Thirlwall to Whewell, September 23, 1834, TC: Add.Ms.a.213/188.

42. William Whewell, "Lecture VIII: The Relation of Church and State," in his *Lectures on Systematic Morality*, 54–61, 56–57.

43. Julius Charles Hare, *Miscellaneous Pamphlets on Some of the Leading Questions Agitated in the Church During the Last Ten Years* (Cambridge, 1855); Hugh James Rose, *Commission and Consequent Duties of the Clergy in a Series of Discourses Preached Before the University of Cambridge in April 1828* (Cambridge, 1828).

44. Thirlwall to Whewell, September 1, 1834, TC: Add.Ms.a.213/177.

45. Jones to Whewell, June 23, 1834, TC: Add.Ms.c.51/171.

46. Thirlwall, *Connop Thirlwall*, 81–82.

47. J. W. Clark and H. C. G. Matthew, "Thirlwall, (Newell) Connop"; and Thirlwall, *Connop Thirlwall*, 109–11.

48. Gibson, *Church, Sate and Society*, 38–39, 133–35.

49. N. Merrill Distad, "Hare, Julius Charles (1795–1855)," *ODNB*.

50. Todhunter, *William Whewell*, 1:64; and Douglas, *Life of Whewell*, 144.

51. The sources for this paragraph on the SPCU are found at TC: C.13.61–63, 66, 75.

52. "A New Song to an Auld Fling" and "Stanzas," 1837, TC: C.13.60.

CHAPTER FIVE: AN ETHICAL ECONOMY

Epigraph: Whewell to Jones, October 29, 1851, in Todhunter, *William Whewell*, 2:369–70.

1. Eric Roll, *A History of Economic Thought*, 5th ed. (London: Faber and Faber, [1938] 1992), 284–85.

2. Gerard M. Koot, *English Historical Economists, 1870–1926: The Rise of Economic History and Neomercantilism* (Cambridge: Cambridge University Press, 1987).

3. Goetz A. Briefs, "The Economic Philosophy of Romanticism," *Journal of the History of Ideas* 2, no. 3 (June 1941).

4. William Whewell, preface to Jones, *Literary Remains*, xi.

5. Richard Jones, *An Essay on the Distribution of Wealth and on Sources of Taxation* (London, 1831), vii.

6. Whewell to Jones, May 24, 1825, TC: Add.Ms.c.51/21.

7. Whewell to Jones, April 24, 1831, in Todhunter, *William Whewell*, 2:118.

8. Whewell to Jones, September 27, 1827, TC: Add.Ms.c.52/15.

9. Whewell to Jones, May 18, 1828, TC: Add.Ms.c.51/52.

10. Jones, *An Essay on the Distribution of Wealth*, x–xi, xxx–xxxii, xxxviii.

11. Whewell to Jones, February 19, 1832, in Todhunter, *William Whewell*, 2:139–42.

12. Whewell to Jones, December 27, 1832, in Todhunter, *William Whewell*, 2:151–52.

13. Whewell to Jones, September 10, 1827, TC: Add.Ms.c.51/41; Hilton, *Mad, Bad and Dangerous People?*, 502–5.

14. Hugh James Rose, *The Farmers and the Clergy: Six Letters to the Farmers of England, on Tithes and Church Property* (London, 1831); [Richard Jones], "On Tithes," in *The British Magazine* (London, 1832), 2:252–58.

15. Whewell to Jones, October 11, 1832, TC: Add.Ms.c.51/142. Whewell hoped the magazine would persuade laymen that the Church was not just for the clergy; see Whewell to Rose, February 3, 1832, in Todhunter, *William Whewell*, 2:138–39.

16. Edward Bromhead to Whewell, April 29, 1833, TC: Add.Ms.a.201/202.

17. Townsend to Whewell, October 7, 1832, TC: Add.Ms.a.213/146. For Marx's use of Jones see especially *Capital: A Critique of Political Economy*, vol. 1, transl. Ben Fowkes (St. Ives, UK: Penguin, [1867] 1990), and vol. 3, trans. David Fernbach (St. Ives, UK: Penguin, [1894] 1991).

18. Jones, *Literary Remains*, 410–11.

19. Henryk Grossman, "The Evolutionist Revolt against Classical Economics: II. In England—James Steuart, Richard Jones, Karl Marx," *Journal of Political Economy* 51, no. 6 (December 1943), 508–13.

20. Whewell, *On the Philosophy of Discovery*, 296.

21. Whewell reiterated Jones's argument in his lectures to the Prince of Wales in 1861; see his *Six Lectures on Political Economy* (Cambridge, 1862), 65–68.

22. Richard Jones, *Peasant Rents: Being the First Half of an Essay on the Distribution of Wealth and on the Sources of Taxation* (London, 1831); this work is usefully summarized in Whewell, *On the Philosophy of Discovery*, 297.

23. [John Ramsay McCulloch], "Review of Richard Jones, *An Essay on the Distribution of Wealth, and on the Sources of Taxation* (London, 1831)," *Edinburgh Review*, August–December 1831, 85.

24. Whewell, "Notes on Political Economy," TC: Add.Ms.c.53[6–11].

25. Anon., "Review: Introductory Lecture on Political Economy, delivered at King's College, London. By the Rev. R. Jones, M.A. London. Murray, 1833," *British Magazine and Monthly Register of Religious and Ecclesiastical Information*, August 1833, 581. The author could well have been Rose, who not only helped found this magazine but also helped edit it.

26. Whewell to Jones, February 20, 1835, TC: Add.Ms.c.51/179.

27. Whewell to Jones, April 12, 1835, TC: Add.Ms.c.51/180. For the volatile Irish question and the future of both the Irish Church and Church of England, see Chadwick, *Victorian Church*, 47–61.

28. Chadwick, *Victorian Church*, 184.

29. Conybeare to Whewell, April 18, 1835, TC: Add.Ms.a.208/93.

30. Snyder explores the overlaps and differences among Jones's, Whewell's, and Malthus's conclusions in *Reforming Philosophy*, chap. 5; and Maas, "'Hard Battle to Fight,'" 149–50. Marx concluded that Jones, as one of a trinity of economic parsons (Malthus and Chalmers being the other two), was far superior; see his *Capital*, 3:375.

31. J. M. Pullen, "Jones, Richard (1790–1855)," *ODNB*.

32. D. C. Moore, "Social Structure, Political Structure, and Public Opinion in Mid-Victorian England," in Robson and Cannon, *Ideas and Institutions of Victorian Britain*, 49; and Snyder, *Philosophical Breakfast Club*, 182–84.

33. Richard Cobden to Joseph Sturge, July 25, 1842, quoted in Jones, *End to Poverty?*, 189.

34. Jones to Whewell, November 11, 1835, TC: Add.Ms.c.52/64.

35. Whewell to Rose, August 12, 1836, TC: R.2.99/42.

36. Jones to Whewell, May 11, 1837, TC: Add.Ms.c.52/68. For Whewell's subsequent celebration, years after Rose had died, of Jones's work on the commutation of the tithes; see his introduction to Jones, *Literary Remains*.

CHAPTER SIX: THE NEW MASTER OF TRINITY

1. Whewell to Hare, December 15, 1840, TC: Add.Ms.a.215/53; and Hare to Whewell, December 17, 1840, TC: Add.Ms.a.206/177. See also Todhunter, *William Whewell*, 1:150. Whewell had dedicated his Bridgewater Treatise to Blomfield in 1837. Whewell had also considered resigning his employment as a tutor in October 1833; see Whewell to Jones, October 21, 1833, in Todhunter, *William Whewell*, 2:170–71. Instead he made a deal with Perry and Thirlwall to dilute his workload; see Whewell to Jones, November 13, 1833, in Todhunter, *William Whewell*, 2:172–73.

2. Although there had been some talk of Wordsworth's son, also named Christopher, succeeding him; see Worsley to Hare, October 13, 1841, in Douglas, *Life of Whewell*, 226.

3. Blomfield to Peel, September 21, 1841, and quoted in Robson and Cannon, "Trinity College," 315. See also Robson and Cannon, "William Whewell," 171.

4. Peacock to Whewell, n.d. (probably October 18, 1841), in Douglas, *Life of Whewell*, 229; and Whewell to Peacock, October 19, 1841, in Todhunter, *William Whewell*, 2:301–2.

5. Helena M. Pycior, "George Peacock and the British Origins of Symbolical Algebra," *Historia Mathematica* 8, no. 1 (February 1981): 24–28. For Peacock's natural history of mathematics, see Kevin Lambert, "A Natural History of Mathematics: George Peacock and the Making of English Algebra," *Isis* 104, no. 2 (2013). This history of mathematical progress had resulted in symbolical algebra as a universal form of human reasoning.

6. Whewell to Hare, October 30, 1841, TC: Add.Ms.a.215/62.

7. Whewell to Hare, October 30, 1841; and Whewell to Hare, March 29, 1843, in Douglas, *Life of Whewell*, 286.

8. Whewell to [Thomas Stephen] Gossett, January 4, 1842, in Douglas, *Life of Whewell*, 255. Thomas Stephen Gosset was a senior Trinity fellow and since 1824 the vicar of Old Windsor.

9. Cooper, *Annals of Cambridge*, 4:621. For Whewell's hostility to Peel, see Hilton, *Mad, Bad and Dangerous People?*, 508–9, 546–58. Barnwell was a sore point for men like Whewell who boiled "the destitute state" of the area down to religious neglect. Indeed, Perry went as far as purchasing Barnwell's church in an attempt to improve the local religious instruction; see Whewell to Hare, December 26, 1836, in Todhunter, *William Whewell*, 2:250.

10. Jones to Whewell, February 17, 1846, TC: Add.Ms.c.52/104; and Whewell to Jones, February 18, 1846, TC: Add.Ms.c.51/249. For Peel and his use of the Irish Famine to repeal the Corn Laws, see, for example, G. Kitson Clark, *The Making of Victorian England* (London: Routledge, 1962), 45–48; and Jonathan Parry, *The Rise and Fall of Liberal Government in Victorian Britain* (New Haven, CT: Yale University Press, 1993), 164–65.

11. Hilton, *Mad, Bad & Dangerous People?*, 556–58.

12. Cooper, *Annals of Cambridge*, 4:680. Tom Thumb or General Tom Thumb as he was more widely known, was really Charles Stratton. He was born in Bridgeport, Connecticut, and was 3.35 feet (99 cm) tall.

13. Whewell to Sedgwick, June 6, 1845, quoted in John Willis Clark, *The Life and Letters of the Reverend Adam Sedgwick* (1890; repr., London: Forgotten Books, 2015), 2:98.

14. Whewell to the Porter, March 16, 1846, TC: 0.15.45/182.

15. Whewell to Jones, June 11, 1846, TC: Add.Ms.c.51/250.

16. Whewell to Hare, June 17, 1847, TC: Add.Ms.a.215/96.

17. Whewell to the Tutors, April 26, 1847, TC: 0.15.45/208.

18. Henry Sidgwick, "Philosophy at Cambridge," *Mind* 1 (1876): 235–46.

19. Yeo, *Defining Science*, 62–69.

20. F. B. Smith, "The Atheist Mission 1840–1900," in Robson and Cannon, *Ideas and Institutions*, 205–35; Christopher Hamlin, *Public Health and Social Justice in the Age of Chadwick: Britain, 1800–1854* (Cambridge: Cambridge University Press, 1998).

21. Thirlwall, *Connop Thirlwall*, 205. Newman was made a cardinal by Pope Leo XIII in February 1879 and canonized as a saint in 2019.

22. Trench to Whewell, December 31, 1846, TC: Add.Ms.a.64/102.

23. Chadwick, *Victorian Church*, 211.

24. Whewell to Prince Albert, March 8, 1847, TC :0.15.45/206.

25. Chadwick, *Victorian Church*, 163–64.

26. William Whewell, letter to the editor, *Cambridge Advertiser and University Herald*, March 3, 1847, 182, and cited in "University of Cambridge Chancellor Election, 1847," *Wikipedia*, accessed at https://en.wikipedia.org/wiki/University_of_Cambridge_Chancellor_election,_1847; Douglas, *Life of Whewell*, 338–43.

27. Douglas, *Life of Whewell*, 301, 371, 407.

28. Whewell to Hare, May 9, 1848, TC: Add.Ms.a.215/97.

29. Richards, *Mathematical Visions*, 41–42, 120.

30. Whewell to Everina E. Affleck, May 12, 1855, in Douglas, *Life of Whewell*, 440.

31. Faye, *Science in Culture*, 54; Snyder, *Reforming Philosophy*, 20, 218–21; and Snyder, *Philosophical Breakfast Club*, 301–2.

32. Trevelyan, *Trinity College*, 96. The first appointment to this chair was a Trinity man and Cambridge Apostle, Sir William Harcourt (1827–1904) in 1868, see Douglas, *Life of Whewell*, 344.

33. Whewell, *Aphorisms Concerning Ideas, Science, and the Language of Science* (London, 1840), 34, and *History of the Inductive Sciences*, 1:8.

34. William Whewell, *On the Material Aids of Education: Being an Inaugural Lecture on the Occasion of the Educational Exhibition of 1854* (London, 1854), 15.

35. Warwick, *Masters of Theory*; and Chapter 8 in this book.

36. Whewell to Hare, March 13, 1842, in Douglas, *Life of Whewell*, 264.

37. Whewell, *On the Material Aids of Education*, 15–16. Italics mine.

38. Whewell, *On the Material Aids of Education*, 18.

CHAPTER SEVEN: GENIUS AND GREAT SCIENTIFIC DISCOVERIES

Epigraph: Whewell, *Astronomy*, 375.

1. Samuel Taylor Coleridge, *Biographia Literaria; or Biographical Sketches of My Literary Life and Opinions* (London, 1817), 2:258, quoted in Prickett, *Coleridge and Wordsworth*, 84. The key place of genius within the so-called Romantic Movement is succinctly explored in Simon Schaffer, "Genius in Romantic Natural Philosophy," in Cunningham and Jardine, *Romanticism and the Sciences*, 82–98.

2. Kant, quoted in Richards, *Romantic Conception of Life*, 70.

3. For the adverse mechanistic and materialistic implications of William Paley, see, for example, Neal C. Gillespie, "Divine Design and the Industrial Revolution: William Paley's Abortive Reform of Natural Theology," *Isis* 81, no. 2 (June 1990); and for a more moderate interpretation that situates Paley's work within the context it was written, see David Burbridge, "William Paley Confronts Erasmus Darwin: Natural Theology and Evolutionism in the Eighteenth Century," *Science and Christian Belief* 10, no. 1 (April 1998). For Paley's prominent place in the Cambridge curriculum for demonstrating the evidence of Christianity, see Fyfe, "The Reception of William Paley's *Natural Theology*."

4. Secord, *Visions of Science*, 10; and Martin J. S. Rudwick, *Earth's Deep History: How It Was Discovered and Why It Matters* (Chicago: University of Chicago Press, 2014), 161.

5. For the historical attempt to bridge religious factions, see especially Brooke, *Science and Religion*.

6. Schaffer, "Genius," 85–86.

7. Julius Charles Hare, "Sermon IV: Power of Faith in Man's Natural Life," in Hare, *Victory of Faith*, 25–36, 31; and Hare, "Sermon VI: Faith in Christ, The Victory That Overcometh the World," in Hare, *Victory of Faith*, 45–57, 51. A similar critical debate on the relationship between natural theology and proof of God was prominent at Oxford; see Corsi, *Science and Religion*, 178–93.

8. Whewell, *Astronomy*, 254–68.

9. Harrison, *Territories of Science*, 28.

10. Harrison, *Territories of Science*, 12.

11. Galileo, quoted in Harrison, *Fall of Man*, 106.

12. Galileo, quoted in Harrison, *Territories of Science*, 77.

13. Harrison, *Fall of Man*, 101.

14. Johannes Kepler, quoted in Harrison, *Fall of Man*, 103.

15. For this quest among medieval theologians, see Harrison, *Territories of Science*, 63–67.

16. John Flamsteed, quoted in George Boole, *The Mathematical Analysis of Logic* (Cambridge, 1847), 13.

17. William J. Ashworth, "'Labour harder than *thrashing*': John Flamsteed, Property and Intellectual Labour in Nineteenth-Century England," in *Flamsteed's Stars: New Perspectives on the Life and Work of the First Astronomer Royal, 1646–1719*, ed. Frances Willmoth (Woodbridge, UK: Boydell Press, 1997), 199–216; Ashworth, "Calculating Eye"; Rebekah Higgitt, "Astronomers against Newton," *Endeavour* 28 (March 2004).

18. For an excellent exposition of this continued emphasis on human computing until its manifestation in electronic computers, see David Alan Grier, *When Computers Were Human* (Princeton, NJ: Princeton University Press, 2005).

19. Francis Baily, *An Account of the Rev. John Flamsteed, the first astronomer Royal; compiled from his own Manuscripts and other authentic Documents, never before published; Part II: John Flamsteed's British Catalogue* (London, 1835), xvii, 370.

20. Simon Schaffer, "Astronomers Mark Time: Discipline and the Personal Equation," *Science in Context* 2, no. 1 (1988); Ashworth, "Calculating Eye."

21. The argument for Flamsteed can be followed in Ashworth, "'Labour harder than *thrashing*,'" 205–8.

22. Whewell to Jones, December 31, 1835, TC: Add.MS.c.51/194. Whewell later chastised De Morgan's work on logic: "You make her the old maid of the family of Truth, who nurses the children of all the sister sciences, but has none herself"; see Whewell to De Morgan, February 14, 1859, in Todhunter, *William Whewell*, 2:417–18.

23. Whewell, *Astronomy*, 315–17. Although by 1790 Kant did not consider Newton a genius like, say, Kepler; for his reasons, see Schaffer, "Genius," 89–90.

24. Julius Charles Hare, "Sermon IV: Power of Faith in Man's Natural Life," in Hare, *Victory of Faith*, 32.

25. Richard Yeo, "Genius, Method, and Morality: Images of Newton in Britain, 1760–1860," *Science in Context* 2, no. 2 (Autumn 1988): 272.

26. Whewell, *On the Philosophy of Discovery*, 361–36.

27. Rob Iliffe, *Priest of Nature: The Religious Worlds of Isaac Newton* (Oxford: Oxford University Press, 2017), 4.

28. Rebekah Higgit, *Recreating Newton: Newtonian Biography and the Making of Nineteenth-Century History of Science* (2007; repr., Pittsburgh: University of Pittsburgh Press, 2020).

29. Baily, *Account of the Rev. John Flamsteed*, xxi–xii.

30. Gascoigne, "Mathematics and Meritocracy," 547–84; and Gascoigne, *Cambridge in the Age of the Enlightenment*, 142–84.

31. Stephen Peter Rigaud to Whewell, January 25, 1836, TC: Add. MS.a.211/79; Stephen Jordan Rigaud, *Memoir of Stephen P. Rigaud* (Oxford, 1883), 11. Riguad's son wrote *A Defence of Halley* (London, 1844).

32. Higgit, *Recreating Newton*.

33. William Whewell, *Newton and Flamsteed: Remarks on an Article in Number CIX of the Quarterly Review* (Cambridge, 1836), 4.

34. Whewell, *Newton and Flamsteed*, 5. Whewell was alert to anyone who dared smear and deflate the name of Newton. See, for example, his defense of Newton against Hegel's negative interpretation, "On Hegel's Criticism of Newton's Principia," *Transactions of the Cambridge Philosophical Society* 51 (1849). Thirlwall described Hegel as "the most impudent of all literary quacks" and was very surprised to see him as having exercised "so great an influence over the mind of Germany"; see Thirlwall to Whewell, October 31, 1849, in Thirlwall, *Letters of Thirlwall*, 195. Whewell claimed Hegel was "equally fanciful and shallow" on every subject, "though he may not be so demonstratively wrong as in the matter of Newton"; Whewell to Hare, October 28, 1848, TC: Add.Ms.a.215/100. Sedgwick also had harsh words to say about Hegel in *A Discourse of the Studies of the University* (Cambridge, 1832; repr., London: Leicester University Press, 1969). For the reception and penetration of Hegelian thought into Britain, see especially Kirk Willis, "The Introduction and Critical Reception of Hegelian Thought in Britain 1830–1900," *Victorian Studies* 32, no. 1 (Autumn 1988).

35. This, of course, is something that continues to plague debates over artificial intelligence. See, for example, the collected essays in Stefano Franchi and Güven Güzeldere, eds., *Mechanical Bodies, Computational Minds: Artificial Intelligence from Automata to Cyborgs* (Cambridge, MA: MIT Press, 2005).

36. Schaffer, "Babbage's Intelligence"; Ashworth, "Calculating Eye"; and Ashworth, "Memory, Efficiency and Symbolic Analysis." For the wider context in which a machine mentality had an impact on scientific work and skill, see Smith and Wise, *Energy and Empire*; Iwan Rhys Morus, "Manufacturing Nature: Science, Technology and Victorian Consumer Culture," *British Journal for the History of Science* 29, no. 4 (1996); Secord, *Visions of Science*, 52–79.

37. Herschel to Whewell, August 6, 1840, TC: Add.Ms.a.207/45.

38. Herschel to Whewell, April 17, 1841, TC: Add.Ms.a.207/46.

39. Whewell, *On the Philosophy of Discovery*, 382.

40. Preyer, "Romantic Tide Reaches Trinity"; Cantor, "Between Rationalism and Romanticism"; Yeo, *Defining Science*, 65–71; John Wettersten, "William

Whewell: Problems of Induction vs. Problems of Rationality," *British Journal for the Philosophy of Science* 45, no. 2 (June 1994): 729. Timothy L. Alborn discusses the divide over the nature of discovery in early nineteenth-century Britain in "The Business of Induction: Industry and Genius in the Language of British Scientific Reform, 1820–1840," *History of Science* 34, no. 1 (March 1996).

41. Whewell to Herschel, April 9, 1836, in Todhunter, *William Whewell*, 2:233; Richards, *Romantic Conception of Life*, 106–11.

42. Augustus De Morgan, quoted in Yeo, *Defining Science*, 142.

43. Whewell was not always complimentary of Macaulay; see, for example, his "On Mr. Macaulay s Praise of Superficial Knowledge," *Fraser's Magazine*, July–December 1849, 171–75, and TC: R.6.14[8]; and Whewell to Forbes, October 14, 1848, in Todhunter, *William Whewell*, 2:348.

44. This was something central to liberal Anglican history; see Forbes, *Liberal Anglican Idea of History*, 63–86, 98–101.

45. Whewell uses Kepler to illustrate his argument in his Bridgewater Treatise *Astronomy*, 305–8, 314; Yeo, *Defining Science*, 157–70.

46. Whewell, *Astronomy*, 303–22.

47. Whewell, quoted in E. W. Strong, "William Whewell and John Stuart Mill: Their Controversy about Scientific Knowledge," *Journal of the History of Ideas* 16, no. 2 (April 1955): 216.

48. Harald T. Walsh, "Whewell and Mill on Induction," *Philosophy of Science* 29, no. 3 (1962); A. W. Heathcote, "William Whewell's Philosophy of Science," *British Journal for the Philosophy of Science* 4, no. 16 (February 1953): 303–4; Richards, *Mathematical Visions*, 28–29; Snyder, *Philosophical Breakfast Club*, 252–53.

49. Abrams, *Mirror and the Lamp*, 115; and Worthen, *Coleridge*, 104.

50. Strong, "Whewell and Mill," 210, 217, 220–23, 227–28; Snyder, *Reforming Philosophy*, 52–53, 104–5; Heathcote, "Whewell's Philosophy of Science," 305; Menachem Fisch, "A Philosopher's Coming of Age: A Study in Esoteric Intellectual History," in Fisch and Schaffer, *William Whewell*, 31–66, 35; Wettersten, "William Whewell," 726. For Baden Powell, see Corsi, *Science and Religion*, 156–177; Snyder, "'Whole Box of Tools,'" 209; Fred Wilson, "The Logic of John Stuart Mill," in Gabbay and Woods, *Handbook of Logic*, 231–32.

51. Richards, *Mathematical Visions*, 38.

52. Whewell to Jones, April 7, 1843, TC: Add.Ms.c.51/227, and reprinted in Todhunter, *William Whewell*, 2:313–14. For Whewell's rant against Comte, see especially his later essay, "Comte and Positivism," *Macmillan's Magazine*, 1866.

53. Jones to Whewell, May 10, 1844, TC: Add.Ms.c.52/92.

54. Mill, quoted in Snyder, *Reforming Philosophy*, 4, 95. Snyder traces Mill's hostility to Whewell throughout her book.

55. Strong, "Whewell and Mill," 225.

56. For Coleridge's idea of history, see, for example, Kooy, "Romanticism."

57. Wettersten, "William Whewell," 725; David B. Wilson, "Herschel and Whewell's Version of Newtonianism," *Journal of the History of Ideas* 35, no. 1 (January–March 1974): 81–82; Snyder, *Reforming Philosophy*, 81–82, 97, 132, 143. For the impact and popularity of Herschel's treatise, see Walter F. Cannon, "John Herschel and the Idea of Science," *Journal of the History of Ideas* 22 (April–June 1961); and Secord, *Visions of Science*, chap. 3.

58. Jones to Whewell, May 13, 1837, TC: Add.Ms.c.52/69.

59. Whewell, quoted in Yeo, "Whewell," n.p.

60. William Whewell, *Aphorisms Concerning Ideas, Science, and the Language of Science* (London, 1840), 22.

61. Richard Whately, *Introduction to Political Economy* (London, 1832), 18, 21; and James Van Evra, "Richard Whately and Logical Theory," in Gabbay and Woods, *Handbook of the History of Logic*, 75–91, 84–85.

62. Ashworth, "Memory, Efficiency and Symbolic Analysis." A year later Whewell condemned Whately's political economy and poor use of language; see Schaffer, "Whewell's Politics of Language," 223–24.

63. Corsi, *Science and Religion*, 152–53.

64. Maas, "'Hard Battle to Fight,'" 155–56.

65. Jones to Whewell, February 24, 1831, TC: Add.Ms.c.52/20.

66. Whewell to Jones, February 25, 1831, TC: Add.Ms.c.51/99; William Whewell, "Modern Science: Inductive Philosophy," *Quarterly Review* 45 (1831).

67. Jones to Whewell, February 25, 1831, TC: Add.Ms.c.52/21.

68. Whewell to Jones, July 15, 1831, TC: Add.Ms.c.51/86.

69. Whewell to Jones, July 23, 1831, TC: Add.Ms.c.51/110, and reprinted in Todhunter, *William Whewell*, 2:124–26.

70. Whewell to Jones, December 21, 1832, TC: Add.Ms.c.51/147.

CHAPTER EIGHT: TRINITY AND MORAL PHILOSOPHY

Epigraphs: William Whewell, *The Elements of Morality, Including Polity*, 3rd ed. (Cambridge, [1841] 1854), 1:22. This quote is taken from the Preface of the second edition (1848) that is reprinted in the third edition. Whewell, *On the Philosophy of Discovery*, 381.

1. Desmond and Moore, *Darwin's Sacred Cause*, 25, 61–63.

2. Whewell to Jones, May 9, 1835, TC: Add.Ms.c.51/182; Hilton, *Mad, Bad and Dangerous People?*, 348.

3. James Mackintosh, *Dissertation Second: Exhibiting a General View of the Progress of Ethical Philosophy with a Preface by William Whewell* (Cambridge, 1835).

4. Whewell to Jones, August 21, 1835, TC: Add.Ms.c.51/187, and reprinted in Todhunter, *William Whewell*, 2:222–23.

5. Hallam to Whewell, January 6, 1836, TC: Add.Ms.a.205/84.

6. Desmond and Moore, *Darwin*, 283; and Richards, *Romantic Conception of Life*, 549.

7. Whewell to Jones, September 22, 1833, TC: Add.Ms.a.215/28.

8. Hare to Whewell, July 13, 1836, TC: Add.Ms.a.206/169.

9. Whewell to Hare, March 13, 1842, TC: Add.Ms.a.215/66.

10. Whewell to his sister, August 11, 1842, in Douglas, *Life of Whewell*, 273; Whewell to Hare, February 19, 1842, TC: Add.Ms.a.215/65; Todhunter, *William Whewell*, 1:120. It seems the only work deriving from the Benthamites that Whewell had time for was John Austin's *The Province of Jurisprudence Determined* (1832); see his "Notes on Bentham 1848–1849," TC: R.6.13[34]. Austin had taught Mill Roman law and legal analysis; see Preyer, "John Stuart Mill," 44.

11. Whewell to Hare, January 7, 1838, in Todhunter, *William Whewell*, 2:268.

12. Hare to Whewell, December 3, 1841, TC: Add.Ms.a.77/131.

13. Whewell to Hare, December 8, 1841, TC: Add.Ms.a.215/64. For Coleridge and Maurice, see Smith, *Gothic Revival*, 153–54, 181–82; and for Maurice and Niebuhr, see Rogerson, *Old Testament Criticism*, 188; and John Rogerson, *The Bible and Criticism in Victorian Britain: Profiles of F. D. Maurice and William Robertson Smith* (Sheffield, UK: Sheffield Academic Press, 1995).

14. Hare to Whewell, March 1, 1845, TC: Add.Ms.a.77/140. I am grateful to Chris Stray for the translation of the Greek word *hysteroproterites*.

15. Whewell to Jones, March 7, 1845, TC: Add.Ms.c.51/236.

16. Whewell, *Elements of Morality*, 1:v. Butler was a staunch opponent of deism, Hobbes, and Locke. He is most well known for his book *Analogy of Religion, Natural and Revealed* (1736).

17. Whewell, *Thoughts on the Study of Mathematics*, 42.

18. Myers to Whewell, July 31, 1845, and Whewell to Myers, September 6, 1845, both in Douglas, *Life of Whewell*, 323, 327.

19. Hare to Whewell, March 31, 1842, in Douglas, *Life of Whewell*, 287; and Hare to Whewell, March 29, 1843, TC: Add.Ms.a.206/180.

20. Douglas, *Life of Whewell*, 275–76, 287.

21. William Whewell, *The Christian's Duty Towards Transgressors: Sermon Preached in the Chapel of the Philanthropic Society, St. Georges Fields, on Sunday, May 16, 1847* (London, 1847), 3–5.

22. Whewell, *Christian's Duty Towards Transgressors*, 8, 21.

23. Whewell, "Sermons," Trinity Chapel, March 26, 1843, TC: R.6.17[35].

24. Whewell, "Sermons," Trinity Chapel, April 29, 1860, TC: R.6.17[31].

25. Cantor, "Between Romanticism and Rationality," 72–78, 82.

26. William Whewell, *On the Foundations of Morals: Four Sermons Preached before the University of Cambridge, November 1837* (New York, 1839), vii. Hare was "deeply honoured" by the dedication and told Whewell that many of the points in his sermons could also be found in the earlier works of the Scottish

preacher Thomas Chalmers; see Hare to Whewell, January 4, 1838, TC: Add. Ms.a.206/171.

27. Whewell, *On the Foundations of Morals*, 18. It is worth comparing this with St. Paul's prior two statements: "18 The wrath of God is being revealed from heaven against all the godlessness and wickedness of people, who suppress the truth by their wickedness, 19 since what may be known about God is plain to them, because God has made it plain to them. 20 For since the creation of the world God's invisible qualities—his eternal power and divine nature—have been clearly seen, being understood from what has been made, so that people are without excuse."

28. Whewell, *On the Foundations of Morals*, 21–23. Here Whewell was chastising Paley's use of a watch to prove his argument for intelligent design, see Paley, *Natural Theology*, especially, the first three chapters.

29. Whewell, *On the Foundations of Morals*, 25–26.

30. Whewell, *On the Foundations of Morals*, 36–37.

31. Whewell, *On the Foundations of Morals*, 50.

32. Whewell, *On the Foundations of Morals*, 53–54.

33. Whewell, *On the Foundations of Morals*, 59–61.

34. Whewell, *On the Foundations of Morals*, v, vii; and Whewell, *Lectures on Systematic Morality*.

35. Hare, *Charges to the Clergy of the Archdeaconry of Lewes*, lx.

36. Whewell to Rose, June 24, 1818, in Todhunter, *William Whewell*, 2:26; Brent, *Liberal Anglican Politics*, 153–57; Corsi, *Science and Religion*, 76–77.

37. Harvey Becher, "William Whewell's Odyssey: From Mathematics to Moral Philosophy," in Fisch and Shaffer, *Whewell: A Composite Portrait*, 19–25.

38. Hare, *Charges to the Clergy of the Archdeaconry of Lewes*, xxx, xlix.

39. Perhaps the most articulate was Julius Charles Hare; see his collection of sermons in Hare, *Victory of Faith*, 26–28, 34, 39.

40. Whewell to Jones, October 31, 1832, TC: Add.Ms.c.51/143.

41. William Whewell, "Sermons," St. Mary's, Cambridge, June 30, 1844, TC: R.6.17[56].

42. Whewell, "Sermons," Trinity Chapel, April 29, 1849, TC: R.6.17[61]. See also his "Sermons," Trinity Chapel, February 17, 1850, TC: R.6.17[64]. For the impact of the European revolutions of 1848 on European thought, including Britain, see Douglas Moggach and Gareth Stedman Jones eds., *The 1848 Revolutions and European Political Thought* (Cambridge: Cambridge University Press, 2020).

43. Connop Thirlwall, "The Excellence of Wisdom," Llandingad Church, December 13, 1849, in Thirlwell, *Essays, Speeches and Sermons of Thirlwall*, 343–45, 347, 349.

44. Whewell, *Of the Plurality of Worlds*, 384–85. This historical debate is masterly told in Michael J. Crowe, *The Extraterrestrial Life Debate 1750–1900:*

The Idea of a Plurality of Worlds from Kant to Lowell (Cambridge: Cambridge University Press, 1988); and "William Whewell, The Plurality of Worlds, and the Modern Solar System," *Journal of Religion and Society* 51, no. 2 (2016).

45. Whewell, *Of the Plurality of Worlds*, 404–5.

CHAPTER NINE: SPACE, TIME, AND THE IDEA OF A CAMBRIDGE EDUCATION

Epigraph: Whewell to George William Lyttelton, 4th Baron Lyttelton, March 10, 1840, in Douglas, *Life of Whewell*, 201. Lyttelton was educated at Eton and Trinity—he was a former pupil of Whewell's and at this time being promoted by his former teacher as the candidate for the post of high steward at the university. The eventual appointment was John Singleton Copley, 1st Baron Lyndhurst. Lyndhurst was also a graduate of Trinity, where he had finished second wrangler.

1. William Whewell, "Personal Notebook," n.d., TC: R.18.168.

2. Hare to Whewell, December 17, 1840, in Douglas, *Life of Whewell*, 209.

3. Trevelyan, *Trinity College*, 96.

4. For Arnold, see Forbes, *Liberal Anglican Idea of History*, 111–16; and Collini, *Matthew Arnold*.

5. Whewell, *Thoughts on the Study of Mathematics*, 42–44, 3.

6. Hare, quoted in Distad, *Guessing at Truth*, 26.

7. Milnes, "Coleridge's Logic," 57–58.

8. Charles Babbage, *Passages from the Life of a Philosopher*, vol. 11 in *The Works of Charles Babbage*, ed. Martin Campbell-Kelly (1864; repr., London, Pickering & Chatto, 1991), 46; and H. W. Buxton, *Memoir of the Life and Labours of the Late Charles Babbage*, ed. Anthony Hyman (Cambridge, MA: MIT Press, 1988), 48.

9. Warwick, *Masters of Theory*.

10. Warwick, *Masters of Theory*, 37–38, 51–52; and Harvey W. Becher, "William Whewell and Cambridge Mathematics," *Historical Studies in the Physical Sciences* 11, no. 1 (1980): 3–25.

11. Whewell, *On the Philosophy of Discovery*, 151.

12. Forbes to Whewell, January 8, 1845, TC: Add.Ms.a.204/60; and Airy to Whewell, March 1, 1845, TC: Add.Ms.a.200/58.

13. Forbes to Whewell, February 17, 1846, TC: Add.Ms.a.204/71.

14. Smith and Wise, *Energy and Empire*, 55–59, 64–65.

15. Whewell to Hare, March 31, 1843, TC: Add.Ms.a.215/68.

16. Whewell, quoted in Warwick, *Masters of Theory*, 103. For the Commissioners Report, see Richards, *Mathematical Visions*, 47–49.

17. Warwick, *Masters of Theory*, 127, 181.

18. Whewell to Hare, August 12, 1845, TC: Add.Ms.a.215/81.

19. William Whewell, *Of a Liberal Education in General, and with Particular Reference to the Leading Studies of the University of Cambridge*. Cambridge, 1848.

20. Whewell to Jones, September 28, 1845, TC: Add.Ms.c.51/247. In any case, the emphasis on geometry overcame the analytical threat; see Richards, *Mathematical Visions*, 126–27.

21. Airy to Whewell, August 25, 1845, TC: Add.Ms.a.200/60.

22. Michael Hoskins, "Astronomers at War: South versus Sheepshanks," *Journal of the History of Astronomy* 20, no. 3 (1989); Doron Swade, "Calculation and Tabulation in the Nineteenth Century: Airy versus Babbage" (PhD diss., University College London, 2003), 228–36.

23. It is interesting to note that as a student Airy had been excited by Babbage's machine and had attempted to make one himself to solve cubic equations; see Whewell to Herschel, October 17, 1822, in Todhunter, *William Whewell*, 2:50–51. This, as Doron Swade speculates, may indicate an element of jealousy in Airy's reasoning; see Swade, "Calculation and Tabulation," 236. For Airy's official reasons to end funding, see Swade, "Calculation and Tabulation," chap. 4; and his *Cogwheel Brain*, chap. 7.

24. Airy to Whewell, October 5, 1845, TC: Add.Ms.a.200/61. The St. John's senior wrangler and biographer of aspects of Whewell's work Isaac Todhunter later agreed with Whewell's and Airy's conclusion regarding analytical mathematics; see Todhunter, *William Whewell*, 158–59.

25. Warwick, *Masters of Theory*, 94–99, 101–2, 107.

26. James David Forbes, *The Danger of Superficial Knowledge: An Introductory Lecture to the Course of Natural Philosophy in the University of Edinburgh* (London, 1849), 60–61.

27. Eric Ashby and Mary Anderson, "Introduction," in Adam Sedgwick, *A Discourse on the Studies of a University Education* (1832; repr., London: Leicester University Press, 1969), 7; and Whewell to Hare, December 25, 1833, in Todhunter, *William Whewell*, 2:174. In fact, Whewell initially told Hare that the sermon "which, though most unskilfully written and delivered, had in it so much of good, that the undergraduates begged him to print it; perhaps in half a year longer he may do so"; Whewell to Hare, February 17, 1833, in Todhunter, *William Whewell*, 2:158–60. By contrast he was not so impressed by Peacock's sermon, which he described as "rather like an article in the Morning Post than a sermon." However, he did rejoice at a dissertation delivered by his student "Birks" (Thomas Rawson Birks): "The philosophy was most profound and consistent, and the views of the nature of morality of the pure and elevated kind, which I hope we shall always hear from our best men here"; Whewell to Hare, December 25, 1833, in Todhunter, *William Whewell*, 2:175. Birks went on to become Knightsbridge Professor of Moral Philosophy and someone Whewell felt was "on the same track of reasoning with myself"; see Whewell to James Stephen, November 14, 1853, in Todhunter, *William Whewell*, 2:396.

28. Whewell to Rose, January 7, 1834, TC: R.2.99/39, and repr. in Todhunter, *William Whewell*, 2:176–77.

29. Sedgwick, *Discourse*, vi–vii, 2–4.

30. Whewell, *On the Philosophy of Discovery*, 381, and *On the Plurality of Worlds*, 382–85.

31. Sedgwick, *Discourse*, 11–12, 17.

32. Whewell to Hare, October 30, 1841, TC: Add.Ms.a.215/62.

33. Sedgwick, *A Discourse*, 23, 29, 30, 32–34, 36, 38. This argument in some ways runs through the recent histories of the subject by Larry Siedentop and Peter Harrison. For the former, see his *Inventing the Individual*; and for the latter, see his *Fall of Man* along with *Territories of Science*.

34. Sedgwick, *Discourse*, 41, 42, 46.

35. Sedgwick, *Discourse*, 49, 53, 56, 57, 64, 77.

36. Whewell, *Aphorisms*, 32.

37. Therefore recent philosophers of science have contrasted Whewell with the late twentieth-century philosophy of science expounded by Thomas Kuhn. For the latter, see his *Structure of Scientific Revolutions* (Chicago: University of Chicago Press, 1962). Clearly Kuhn's notion of paradigm shifts that are incommensurable with former views of knowledge is the opposite to Whewell's view. For a useful precursor to this debate see Forbes, *Liberal Anglican Idea of History*.

38. Jones to Whewell, May 17, 1843, TC: Add.Ms.c.52/81.

39. Jon Agar, *The Government Machine: A Revolutionary History of the Computer* (Cambridge, MA: MIT Press, 2003), 56–58, 64–66. The shift to a mechanistic system of coaching is told in Andrew Warwick's *Masters of Theory* and is discussed in the Conclusion.

40. William Whewell, "Comte and Positivism," *Macmillan's Magazine*, 1866, 353, 355.

41. Hare to Whewell, November 9, 1849, TC: Add.Ms.a.64/117.

42. Todhunter, *William Whewell*, 1:263.

43. Whewel to Rose, June 24, 1818, TC: R.2.99/10; Cranston, *Romantic Movement*, 31; Robson and Cannon, "William Whewell," 177. For the whole question of a Gothic past, see Smith, *Gothic Bequest*; and for William Wordsworth, see Prickett, *Coleridge and Wordsworth*, 5.

44. William Whewell, *On German Churches. A New Edition. To which is Now Added Notes Written during an Architectural Tour of Picardy and Normandy* (Cambridge, [1830] 1835); Quinn, "William Whewell's Philosophy of Architecture," 13–15.

45. Yanni, "On Nature and Nomenclature," 206; and Buchanan, *Robert Willis*, 76.

46. Whewell, quoted in Buchanan, *Robert Willis*, 78.

47. Buchanan, *Robert Willis*, 79–81.

48. Whewell, *On German Churches*; Yanni, "On Nature and Nomenclature," 209. The 1835 edition of Whewell's 1830 book on German churches has an add-

ed account of an architectural tour of Picardy and Normandy that he conducted with Rickman.

49. Whewell to his sister, August 20, 1832; Whewell to his aunt, September 2, 1832; and Whewell to his sister, June 5, 1833, all in Douglas, *Life of Whewell*, 146–47, 153.

50. Yanni, "On Nature and Nomenclature," 211; and Buchanan, *Robert Willis*, 149.

51. Whewell to Jones, July 31, 1829, in Todhunter, *William Whewell*, 2:99.

52. Douglas, *Life of Whewell*, 128–30; 365.

53. Buchanan, *Robert Willis*, 85–89, 105.

54. Whewell to his sister, August 19, 1849; and Whewell to Kate Malcolm, May 22, 1857, both in Douglas, *Life of Whewell*, 365, 493.

55. Beresford Hope to Whewell, October 25, 1841, in Douglas, *Life of Whewell*, 251.

56. Buchanan, *Robert Willis*, 367.

57. Trevelyan, *Trinity College*, 97–98; Douglas, *Life of Whewell*, 517–18; Schaffer, "Whewell's Politics of Language," 216. For Wordsworth and New Court, see Robson and Cannon, "Trinity College," 329.

58. Douglas, *Life of Whewell*, 453–54; Todhunter, *William Whewell*, 1:216; Buchanan, *Robert Willis*, 39.

59. Robson and Cannon, "William Whewell," 173.

60. Roger Wolfe, "'Quite a Gem': An Account of the Former Mortuary Chapel at Mill Road Cemetery, Cambridge," *Cambridge Antiquarian Society* 84 (1996). I am grateful to James Moore for this reference.

61. Douglas, *Life of Whewell*, 556.

62. Richard Whately, *Introduction to Political Economy* (London, 1832), 18.

63. David de Giustino, "Finding an Archbishop: The Whigs and Richard Whately in 1831," *Church History* 64, no. 2 (June 1995). For more on Whately and his theological and political context, see Boyd Hilton, "Whiggery, Religion and Social Reform: The Case of Lord Morpeth," *Historical Journal* 37, no. 4 (December 1994): 833–34, 856; and Corsi, *Science and Religion*, 97.

64. Snyder, *Reforming Philosophy*, 34–36. For Whately's attempt to reform the education curriculum at Oxford University, see Corsi, *Science and Religion*, 94–95.

65. Maas, "'Hard Battle to Fight,'" 159–60.

66. Whewell, "Mathematical Notes," n.d. (probably 1845), TC: R.6.14[29]. He made related remarks in his *Liberal Education*, 41.

67. Charles Babbage, *The Economy of Machinery and Manufactures* (1832), vol. 8 of *Works of Charles Babbage*, 379.

68. Robert Wilberforce, *Considerations Respecting the Most Effected Means of Encouraging Mathematics* (1830) and quoted in Corsi, *Science and Religion*, 113.

69. This is arguably not so far from E. P. Thompson's later interpretation of the eighteenth century; see especially his collected essays in *Customs in Com-*

mon: Studies in Popular Culture (London: Merlin Press, 1991); and this book's Epilogue.

70. Whewell, *Thoughts on the Study of Mathematics*, 28, 40–46; and Smith and Wise, *Energy and Empire*, 62–63.

71. Hawtrey to Whewell, May 1, 1850, TC: Add.Ms.a.206/18.

72. Hawkins to Whewell, November 5, 1850, TC: Add.Ms.a.206/46.

73. Whewell to the vice-chancellor, May 12, 1851, TC: Add.Ms.a.55/28.

74. Whewell to Jones, August 13, 1851, TC: Add.Ms.c.51.263; and Jones to Whewell, August 15, 1851, TC: Add.Ms.c.52/150.

75. Whewell to Jones, February 21, 1852, TC: Add.Ms.c.51/267.

76. Jones to Whewell, March 24, 1852, TC: Add.Ms.c.52/161.

77. Jones to Whewell, November 15, 1852, TC: Add.Ms.c.52/173; and Jones to Whewell, November 24, 1852, TC: Add.Ms.c.52/174.

78. Jones to Whewell, January 30, 1853, TC: Add.Ms.c.52/160.

79. Milnes to Whewell, March 4, 1853, TC: Add.Ms.a.206/135. Milnes had left Peel's conservative party after the latter passed the Corn Laws in 1846, and in 1863 Lord Palmerston awarded him a peerage, making him Baron Milnes of Great Houghton.

80. W. Garrard to Whewell, April 18, 1853, TC: Add.Ms.a.80/112. This could be Henry Lionel Dymoke of the Dymoke family of Scrivelsby Manor in the parish of Horncastle in Lincolnshire. The family holds the feudal hereditary office of king's champion.

81. Oliver Brose, "F. D. Maurice and the Victorian Crisis of Belief," *Victorian Studies* 3, no. 3 (March 1960): 230.

82. Distad, *Guessing at Truth*, 174–81. For Mill's relationship to Sterling, Thirlwall, and Hare, see Preyer, "John Stuart Mill"; and for Thomas Carlyle, see, for example, Alexander Jordan, "Thomas Carlyle and Political Economy: The 'Dismal Science' in Context," *English Historical Review* 132, no. 555 (April 2017).

83. Hare to Whewell, December 20, 1847, TC: Add.Ms.a.77/161. The result was a two-volume book: John Sterling, *Essays and Tales, by John Sterling. Collected and Edited, with a Memoir of His Life by Julius Charles Hare*, 2 vols. London, 1848.

84. Thomas Carlyle, *The Life of John Sterling*. London, 1851.

85. Thirlwall, *Connop Thirlwall*, 140–42.

86. Whewell, quoted in Schaffer, "Whewell's Politics of Language," 207.

87. Jones to Whewell, March 20, 1848, TC: Add.Ms.c.52/116.

88. Distad, *Guessing at Truth*, 184–85. For Thirlwall's possible contribution, see *Politics of the People: Numbers 1–17* (1848; repr., New York: Augustus M. Kelley, 1971), appendix.

89. Hare to Whewell, April 24, 1848, TC: Add.Ms.a.206/184.

90. Hare to Whewell, November 26, 1853, TC: Add.Ms.a.206/186; and Whewell to his wife, November 28, 1853, in Douglas, *Life of Whewell*, 432. Whewell had long abandoned Sterling and Maurice for their religious and political views; see, for example, Whewell to Hare, October 15, 1838, and December 6, 1843, in Todhunter, *William Whewell*, 2:270–71, 320. See also Brose, "F. D. Maurice," 232–34; and Altholz, *Anatomy of a Controversy*, 6.

91. Hare, quoted in Distad, *Guessing at Truth*, 147.

92. See, for example, Robert Verkaik, *Posh Boys: How the English Public Schools Run Britain* (London: Oneworld Publications, 2018); and Francis Green and David Kynaston, *Engines of Privilege: Britain's Private School Problem* (London: Bloomsbury Publishing, 2019). The aim of public (i.e., private) schools, as the Rugby- and Balliol-educated Thomas H. Green long ago concluded, was to put as much space between the privileged classes and the nation; see Richter, *Politics of Conscience*, 95. This was a point reiterated later by the likewise Rugby- and Balliol-schooled Richard Tawney; see Carter, *T. H. Green*, 178.

93. Thirlwall, "English Education for the Middle Classes," 373, 375–76, 380–82. For the widespread view of this in Victorian Britain, see, for example, Brian Stanley, "'Commerce and Christianity': Providence Theory, the Missionary Movement, and the Imperialism of Free Trade, 1842–1860," *Historical Journal* 26, no. 1 (March 1983).

94. Connop Thirlwall, "The Spirit of Truth, The Holy Spirit," University of Cambridge, May 16, 1869, in Thirlwall, *Essays, Speeches and Sermons of Thirlwall*, 387–400, 393.

95. Whewell, quoted in Schaffer, "Whewell's Politics of Language," 206.

96. Thirlwall to William Henry Lyttelton, January 22, 1867, in Thirlwall, *Letters of Thirlwall*, 255–56.

CONCLUSION: THE LAST DAYS OF THE TRINITY CIRCLE

Epigraphs: William Whewell, "Sermons," Bottisham, May 19, 1853, TC: R.6.17[75]. James J. Sylvester, "Presidential Address: Mathematics and Physics Section," *Report of the Thirty-ninth Meeting of the BAAS held at Exeter in August 1869* (London, 1870), 262, and quoted in Richards, *Mathematical Visions*, 164.

1. Richard Hurrell Froude to the Reverend John Frederick Christie, Italy, April 15, 1833, in Louise Imogen Guiney, *Hurrell Froude: Memoranda and Comments* (London: Methuen, 1904), 103. I am grateful to Simon Schaffer for this quote.

2. The best work on Coleridge's quest for a National Church is still Sanders, *Coleridge and the Broad Church*; and Forbes, *Liberal Anglican Idea of History*.

3. Hall, "What Do Victorian Churches Mean?," 90.

4. Philip S. Bagwell, *The Transport Revolution from 1770* (London: B. T. Batsford, 1974), 125.

5. Whewell to Jones, July 5,1850, TC: Add.Ms.c.51/257.

6. Whewell to Jones, December 12, 1854, TC: Add.Ms.c.51/280.

7. Whewell to James David Forbes, January 4, 1864, and Whewell to Rev. D. Brown, October 26, 1863, both in Todhunter, *William Whewell,* 2:433–35. Richards has shown how the demise of traditional geometry and slow rise of evolutionary theory impacted theology in her *Mathematical Visions,* 104–7. For the debate over the age and history of the earth, see Smith and Wise, *Energy and Empire,* 579–611; and Rudwick, *Earth's Deep History,* esp. chap. 7.

8. Hooker to Whewell, April 20, 1857, TC: Add.Ms.a.206/126.

9. For a succinct overview of the difference between Owen's archetype and that of, say, Johann Wolfgang von Goethe and Darwin's use of the theory, see Richards, *Romantic Conception of Life,* 517, 528–33.

10. Desmond, *Politics of Evolution,* 275; Desmond and Moore, *Darwin,* 368; Hilton, *Mad, Bad and Dangerous People?,* 459–60. For Whewell's celebration of Owen, see, for example, his *Plurality of Worlds,* 334–37, 376–78.

11. Owen took a very Whewellian approach to anatomy; see Adrian Desmond, *Archetypes and Ancestors: Palaeontology in Victorian London 1850–1875* (London: Blond & Briggs, 1982), 42–48; Richards, *Romantic Conception of Life,* 10; Phillip R. Sloan, "Whewell's Philosophy of Discovery and the Archetype of the Vertebrate Skeleton: The Role of German Philosophy of Science in Richard Owen's Biology," *Annals of Science* 60, no. 1 (January 2003). For Whewell's acknowledged debt to Schelling's work, see, for example, the preface to his *Philosophy of the Inductive Sciences.*

12. Whewell to his sister, April 12, 1838, in Douglas, *Life of Whewell,* 191.

13. Desmond, *Archetypes and Ancestors,* 33.

14. William Whewell, "Of Certain Analogies between Architecture and the Other Fine Arts," *Royal Institute of British Architects,* Session 1862–1863, 175–181, quoted in Yanni, "On Nature and Nomenclature," 213–14; and Quinn, "William Whewell's Philosophy of Architecture," 16.

15. Whewell, *Architectural Notes on German Churches,* xii, 4.

16. Desmond, *Archetypes and Ancestors,* 43; Yanni, "On Nature and Nomenclature," 216; Hall, "What Do Victorian Churches Mean?," 82.

17. Desmond and Moore, *Darwin,* 34.

18. Richards, *Romantic Conception of Life,* 10, 516. For the use of Humboldtian Romanticism in Darwin's work, see 518–21, 526.

19. Secord, *Victorian Sensation,* 227–31; Corsi, *Science and Religion,* 229–85, Desmond, *Politics of Evolution,* 176–80; and Desmond and Moore, *Darwin,* 320–23, 487–88.

20. Desmond, *Archetypes and Ancestors,* 55; Adrian Desmond, *Huxley: The Devil's Disciple* (London: Michael Joseph, 1994); Paul White, *Thomas Huxley: Making the "Man of Science"* (Cambridge: Cambridge University Press, 2002).

21. Lyell to Whewell, March 7, 1837, TC: Add.Ms.a.208/128. For philosophical differences between Whewell and Herschel toward Lyell's work, see, for example, Michael Ruse, "Charles Lyell and the Philosophers of Science," *British Journal for the History of Science* 9, no. 4 (December 1976); and Secord, *Visions of Science*, chap. 5.

22. Secord, *Visions of Science*, 150–51.

23. Whewell to Lyell, June 5, 1863, in Todhunter, *William Whewell*, 2:431–32.

24. Lyell to Babbage, May 1832, in *Life, Letters, and Journals of Sir Charles Lyell*, ed. K. Lyell (London, 1881), 2:9–10.

25. Henslow to Whewell, August 16, 1847, TC: Add.Ms.a.206/68.

26. Darwin, quoted in Desmond and Moore, *Darwin*, 236.

27. Burrow, "Uses of Philology," 203; Foster, Clark, and York, *Critique of Intelligent Design*, 102, 123.

28. Preyer, "Language of Discovery," 126.

29. For Sedgwick to Darwin and the *Spectator* quote, see Ashby and Anderson, "Introduction," 15.

30. Desmond and Moore, *Darwin*, 239, 378. For the ongoing debate, compare, for example, Phillip E. Johnson, *The Wedge of Truth: Splitting the Foundations of Naturalism* (Downers Grove, IL: Inter-Varsity Press, 2000), with Foster, Clarke, and York, *Critique of Intelligent Design*.

31. Whewell, "Sermons," Trinity Chapel, October 22, 1865, Colos. [Colossians] 111:2, TC: R.6.17[47]. For a similar message, see also his sermons of November 3, 1861, TC: R.6.17[36], and June 24, 1862, TC: R.6.17[39].

32. Siedentop, *Inventing the Individual*, 64.

33. Whewell, "Sermons," May 27, 1861, Trinity Chapel, TC: R.6.17[34]. See also, for example, Maurice in Brose, "F. D. Maurice," 235; and Sanders, "Coleridge, Maurice and the Church Universal," 43.

34. Maurice, quoted in Sanders, "Coleridge, Maurice and the Church Universal," 44.

35. Whewell, "Sermons," April 26, 1863, Trinity Chapel, TC: R.6.17[41].

36. Whewell, "Sermons," April 26, 1863. See also his *On the Philosophy of Discovery*, 354–75.

37. Whewell to Lyell, May 14, 1863, in Todhunter, *William Whewell*, 2:430.

38. Whewell, "Sermons," April 26, 1863.

39. Whewell, "Sermons," April 26, 1863. See also two further sermons Whewell gave on related subjects in Trinity Chapel: March 6, 1864, TC: R.6.17[42], and August 23, 1864, TC: R.6.17[44].

40. Whewell, "Sermons," Trinity Chapel, February 12, 1865, TC: R.6.17[45].

41. Whewell, "Sermons," Trinity Chapel, February 12, 1865. In 1859 Whewell published his translation of Plato's *Dialogues* in *The Platonic Dia-*

logues for English Readers (Cambridge, 1859), and translated Plato's *Republic* in 1861.

42. Whewell, "Sermons," Trinity Chapel, February 12, 1865.

43. Thirlwall, *Connop Thirlwall*, 195–214, 225.

44. For English awareness of German critical approaches, see Rogerson, *Old Testament Criticism*, 158–79, 209. John W. Parker, ed., *Essays and Reviews* (London, 1860). The reception to the book is best told in Altholz, *Anatomy of a Controversy*.

45. Blomfield, quoted in Kirk Willis, "The Introduction and Critical Reception of Hegelian Thought in Britain 1830–1900," *Victorian Studies* 32, no. 1 (Autumn 1988), 95.

46. Thirlwall, *Connop Thirlwall*, 225–32; Josef L. Altholz, "The Mind of Victorian Orthodoxy: Anglican Responses to 'Essays and Reviews,' 1860–1864," *Church History* 51, no. 2 (June 1982). Probably after being elevated to bishop of Exeter in 1869, Temple withdrew his essay from the book the following year. In 1897 he was rewarded and appointed archbishop of Canterbury. For the prominent role of Pattison in instigating the book, especially his attempt to reconcile theism and science, see Mark Francis, "The Origins of Essays and Reviews: An Interpretation of Mark Pattison in the 1850s," *Historical Journal* 17, no. 4 (December 1974).

47. Rogerson, *Old Testament Criticism*, 217.

48. Thirlwall, quoted in Thirlwall, *Connop Thirlwall*, 233; Brose, "F. D. Maurice," 228.

49. In many ways this demarcation lies at the heart of the recent work of Siedentop and Harrison. For the former, see his *Inventing the Individual*; and for the latter, his *Territories of Science*.

50. Frederick Denison Maurice, *Tracts for Priests and People*, 2nd ed. (London, 1862), 80, and quoted in Brose, "F. D. Maurice," 247.

51. Desmond and Moore, *Darwin*, 500.

52. Altholz, "Mind of Victorian Orthodoxy," 188.

53. Josef L. Altholz, "A Tale of Two Controversies: Darwinism in the Debate over 'Essays and Reviews,'" *Church History* 63, no. 1 (March 1994).

54. Bishop Wilberforce, quoted in Harvey Hill, "Religion and the University: The Controversy over *Essays and Reviews* at Oxford," *Journal of the American Academy of Religion* 73, no. 1 (March 2005): 199; Altholz, *Anatomy of a Controversy*, 21, 52–53, 60, 106–7; and Altholz, "Mind of Victorian Orthodoxy," 195–96; Thirlwall, *Connop Thirlwall*, 228–33. Altholz describes the buildup to the prosecution, the trial, appeal, and subsequent acquittal of the book in his *Anatomy of a Controversy*.. For Powell's reference to Darwin, see Altholz, "Tale of Two Controversies," 52.

55. Charles Bullock, Sir William Denison, and the 7th Earl of Shaftesbury are quoted in Altholz, "Tale of Two Controversies," 54–56.

56. W. H. Brock and R. M. Macleod, "The Scientists' Declaration: Reflexions on Science and Belief in the Wake of 'Essays and Reviews,' 1864–5," *British Society for the History of Science* 9, no. 1 (1976).

57. Hill, "Religion and the University."

58. Hill, "Religion and the University," 184–91. A key resource and inspiration to many of the writers in *Essays and Reviews* was Max Muller, deputy Taylorian Professor of Modern European Language at Oxford. His broad theological views were probably what blocked him from being appointed Boden Professor of Sanskrit in 1860. For many during this era, German professors were represented as the academic model for English professors to emulate, while for conservative opponents—clearly including Thirlwall by this time—this represented infidelity, 195–97.

59. Sheehan, *Enlightenment Bible*, 251.

60. Thirlwall, *Connop Thirlwall*, 259–60.

EPILOGUE: THE DEMISE OF GOD AND THE RISE OF MACHINE INTELLIGENCE

1. Richter, *Politics of Conscience*, 125–31.

2. Green, quoted in David G. Ritchie, *The Principles of State Interference: Four Essays on the Political Philosophy of Mr. Herbert Spencer, J. S. Mill and T. H. Green* (London, 1891), 142.

3. Richter, *Politics of Conscience*, 125.

4. Mackinnon, "Some Aspects of the Treatment of Christianity."

5. Richter, *Politics of Conscience*, 173, 177.

6. Carter, *T. H. Green*, 23, 55. For Tawney and his role in British left politics, see also Ben Jackson, *Equality and the British Left: A Study in Progressive Political Thought, 1900–64* (Manchester: Manchester University Press, 2007); Tim Rogan, *The Moral Economists: R. H. Tawney, Karl Polanyi, E. P. Thompson, and the Critique of Capitalism* (Princeton, NJ: Princeton University Press, 2017); and James Kirby, "R. H. Tawney and Christian Social Teaching: Religion and the Rise of Capitalism Reconsidered," *English Historical Review* 131, no. 551 (August 2016).

7. Richter, *Politics of Conscience*, 101–2.

8. Rogan, *Moral Economists*, 52.

9. For different reasons this is also the more recent conclusion of Larry Siedentop in his book *Inventing the Individual*. He concludes that our failure to understand the intrinsic historical relationship between liberal secularism and Christianity greatly distorts our understanding: "In the Western world today, it contributes to two temptations, to what might be called 'liberal heresies.' The first is the temptation to reduce liberalism to the endorsement of market economics, the satisfaction of current wants or preferences without worrying much about the formation of those wants or preferences. In so doing, it narrows the claims of justice. This temptation reduces liberalism to a crude form of utilitar-

ianism. The second temptation is best described as 'individualism,' the retreat into a private sphere of family and friends at the expense of civic spirit and political participation" (363). Similar conclusions can be found in Kirby, "*R. H. Tawney*."

10. Richter, *Politics of Conscience*, 130–31. Perhaps the most recent exponent of the inequalities of modern capitalism is Joseph E. Stiglitz, *The Price of Inequality* (London: W. W. Norton, [2012] 2013); and Thomas Piketty, *Capital in the Twenty-First Century* (Cambridge, MA: Harvard University Press, 2014).

11. Toynbee provides a short and clear exposition of this view toward the Church and state in an address at Balliol in the spring of 1879, "The Ideal Relation of Church and State"; see Arnold Toynbee, *Toynbee's Industrial Revolution: A Reprint of Lectures on the Industrial Revolution in England, Popular Addresses, Notes and Other Fragments* (1884; repr., Newton Abbot, UK: David & Charles Reprints, 1965), 231–39. Interestingly, two of Green's uncles, on his mother's side, went to both Rugby and Trinity where they thrived studying classics; see Richter, *Politics of Conscience*, 42.

12. Holland, quoted in Carter, *T. H. Green*, 63.

13. Toynbee, quoted in Carter, *T. H. Green*, 72. For Toynbee, see especially Daniel C. S. Wilson, "Arnold Toynbee and the Industrial Revolution: The Science of History, Political Economy and the Machine Past," *History & Memory* 26, no. 2 (Fall/Winter 2014).

14. Such potential has now been realized; see, for example, Wendy Brown, *Undoing the Demos: Neoliberalism's Stealth Revolution* (New York: Zone Books, 2015).

15. Arnold Toynbee, "The Ideal Relation of Church and State," notes of an address delivered at a private meeting in Balliol College, Spring 1879, in *Lectures on the Industrial Revolution*, 251.

16. The impact of Green and British political idealism on Tawney is best told in Carter, *T. H. Green*, 163–85. Followers of Green who were particularly important to Tawney were Edward Caird, Holland, and especially Green's student, Charles Gore. This should be read in tandem with Rogan, *Moral Economists*, 40–41; and Kirby, "R. H. Tawney."

17. Tawney, quoted in Carter, *T. H. Green*, 172.

18. Carter, *T. H. Green*, 173; and Rogan, *Moral Economists*, 37.

19. Toynbee, *Toynbee's Industrial Revolution*, 7–9, 24.

20. Rogan, *Moral Economists*, 8, 56.

21. Rogan, *Moral Economists*, 40–42, 56, 166; and Hirschman, *Passions and the Interests*, 132–35.

22. Thompson, *Customs in Common*, 44–46, 50.

23. Harrison, *Territories of Science*, 159–64, 169.

24. Alan Macfarlane, *F. W. Maitland and the Making of the Modern World* (Cambridge: CreateSpace Independent Publishing Platform, 2018).

25. Gerard M. Koot, "Cunningham, William (1849–1919)," *ODNB*.

26. Koot, "Cunningham," and *English Historical Economics*, 139, 147, 153.

27. Hirschman, *Passions and the Interests*, 38–39, 48–66. For the recent backlash to this, see William Davies, *Nervous States: How Feeling Took over the World* (London: Jonathan Cape, 2018).

28. Daniel Stedman Jones, *Masters of the Universe: Hayek, Friedman, and the Birth of Neoliberal Politics* (Princeton, NJ: Princeton University Press, 2012), 59.

29. Naomi Beck, *Hayek and the Evolution of Capitalism* (Chicago: University of Chicago Press, 2018), 1–9, 16, 23, 28, 47.

30. Quinn Slobodian, *Globalists: The End of Empire and the Birth of Neoliberalism* (Cambridge, MA: Harvard University Press, 2018). For more on the relationship between markets, competition, and knowledge, see Bruce Caldwell, "F. A. Hayek and the Economic Calculus," *History of Political Economy* 48, no. 1 (2016).

31. Maas, "'Hard Battle to Fight,'" 157.

32. Beck, *Hayek*, 24–28.

33. Harrison, *Fall of Man*, 254–55. This is a point celebrated in Mayr, *Authority, Liberty and Automatic Machinery*, 164–65.

34. Slobodian, *Globalists*, 226.

35. Friederich Hayek, *The Sensory Order: An Inquiry into the Foundations of Theoretical Psychology* (Chicago: University of Chicago Press, 1952), 167, and quoted in Beck, *Hayek*, 55.

36. Consider, for example, the historiography purporting to underpin Britain's Industrial Revolution; see especially Joel Mokyr, *The Enlightened Economy: Britain and the Industrial Revolution, 1700–1850* (New Haven, CT: Yale University Press, 2011), and *A Culture of Growth: The Origins of the Modern Economy* (Princeton, NJ: Princeton University Press, 2018).

37. Beck, *Hayek*, 85, 93. For a recent look at the way so-called competition and accompanying emphasis on calculation penetrated our institutions and subsequently its recent erosion, see, for example, Davies, *Limits of Neoliberalism*.

38. Davies, *Limits of Neoliberalism*. As Davies neatly sums up, "It is economic calculation that spreads into all walks of life under neoliberalism, and not markets as such" (xix). One could do a great deal worse than rereading Theodor W. Adorno and Max Horkheimer, *Dialectic of Enlightenment*, which was published the same year as Hayek's *Road to Serfdom* in 1944. For a nuanced reading of the term *neoliberalism*, see Ben Jackson, "Currents of Neo-Liberalism: British Political Ideologies and the New Right, c. 1955–1979," *English Historical Review* 131, no. 551 (August 2016). The neoliberal financialization of everything is best described in Brown, *Undoing the Demos*.

39. For the emphasis on "cultural entrepreneurs," see Mokyr, *Culture of Growth*; and McCloskey, *Bourgeois Dignity*.

40. Verkaik, *Posh Boys*; and Green and Kynaston, *Engines of Privilege*. For the history of those schools and institutions that continue to produce those in power, see especially Patrick Joyce, *The State of Freedom: A Social History of the British State* (Cambridge: Cambridge University Press, 2013), 229–307.

41. Agar, *Government Machine*. Agar's argument that the history of the computer in Britain cannot be told without the history of the state's bureaucratic evolution is convincing and well made. These institutions include both the civil and military sections of the state. For the history of the machine as a metaphor, see Mayr, *Authority, Liberty and Automatic Machinery*. A useful overview of the association between the human mind and mechanization from early history to 2005 is told in Franchi and Güzeldere, *Mechanical Bodies*, 1–149.

42. For a useful engagement in this debate, see, for example, Franchi and Güzeldere, *Mechanical Bodies*; and Stefano Franchi and Francesco Bianchini, eds., *The Search for a Theory of Cognition in Early Mechanisations and New Ideas* (Amsterdam: Rodopi, 2011).

43. George Zarkadakis, *In Our Own Image: Saviour or Destroyer? The History and Future of Artificial Intelligence* (New York: Pegasus Books, 2015), 305, 317.

44. A good starting point for this debate is Graham Button, Jeff Coulter, John R. E. Lee, and Wes Sharrock, *Computers, Minds and Conduct* (Cambridge: Polity Press, 1995); and more recently Harry Collins, *Artificial Intelligence: Against Humanity's Surrender to Computers* (Cambridge: Polity Press, 2018).

BIBLIOGRAPHY

ARCHIVES

William Whewell Archive, Trinity College (TC).
John Herschel Correspondence, St. John's College.
Herschel Papers, Harry Ransom Center, University of Texas at Austin.

PRIMARY SOURCES

Airy, George, and William Whewell. *Account of Experiments made at Dolcoath Mine, I Cornwall, in 1826 and 1828, for the purpose of determining the Density of the Earth.* Cambridge, 1828.

Anon. "Review of *A Treatise on Plane and Spherical Trigonometry.* By John Bonnycastle." *Eclectic Review,* January 1808, 52–59.

Anon. "A Short Memoir of the Rev. Robert Edward Garnham." *Monthly Repository of Theology and General Literature,* January 1815, 13–15.

Anon. "Review: Introductory Lecture on Political Economy, delivered at King's College, London. By the Rev. R. Jones, M.A. London. Murray, 1833." *British Magazine and Monthly Register of Religious and Ecclesiastical Information,* August 1833, 580–81.

Anon. "A New Song to an Auld Fling." 1837. TC: C.13.60.

Anon. "Stanzas." 1837. TC: C.13.60.

Babbage, Charles. "On the Influence of Signs in Mathematical Reasoning." *Transactions of the Cambridge Philosophical Society* 2 (1827, but read on December 16, 1821). In *The Works of Charles Babbage,* 1:371–408.

Babbage, Charles. *The Economy of Machinery and Manufactures.* London, 1832. Vol. 8 of *The Works of Charles Babbage.*

Babbage, Charles. The *Ninth Bridgewater Treatise: A Fragment.* London, 1837. Vol. 9 of *The Works of Charles Babbage.*

Babbage, Charles. *Passages from the Life of a Philosopher.* London, 1864. Vol. 11 of *The Works of Charles Babbage.*

Babbage, Charles. *The Works of Charles Babbage.* Edited by Martin Campbell-Kelly. 11 vols. London: Pickering & Chatto, 1989.

Baily, Francis. *An Account of the Rev. John Flamsteed, the first astronomer Royal; compiled from his own Manuscripts and other authentic Documents, never before published; Part II: John Flamsteed's British Catalogue*. London, 1835.

Bentham, Jeremy. *Church-of-Englandism and Its Catechism Examined*. London, 1818.

Boole, George. *The Mathematical Analysis of Logic*. Cambridge, 1847.

Boole, George. *An Investigation of the Laws of Thought, on which are Founded the Mathematical Theories of Logic and Probabilities*. London, 1854.

Buxton, H. W. *Memoir of the Life and Labours of the Late Charles Babbage*. Edited by Anthony Hyman. Cambridge, MA: MIT Press, 1988.

Church, Dean. *The Oxford Movement: Twelve Years 1833–1845*. London, 1897.

Clark, John Willis. *The Life and Letters of the Reverend Adam Sedgwick*. 2 vols. 1890. Reprint, London: Forgotten Books, 2015.

Clark, W. G. "William Whewell: In Memoriam." *Macmillan's Magazine*, 1866, 545–52.

Cobbett, William. *The Parson and the Boy; Or, The Fire-Shovel Hats in a Bustle: Being an Account of the Trial of the Reverend THOMAS JEPHSON, Doctor of Divinity, and Tutor at Cambridge, on a Charge of attempting to commit an Unnatural Crime on JAMES WELSH, a labourer*. Cambridge, 1823.

Coleridge, Samuel Taylor. *Biographia Literaria; or Biographical Sketches of My Literary Life and Opinions*. 2 vols. London, 1817.

Condillac, Étienne Bonnot de. *Logic, or the First Development of the Art of Thinking* (Paris, 1780). In *Philosophical Writings of Étienne Bonnot de Condillac*, translated by Franklin Philip with Harlan Lane, 343–418. London: Lawrence Erlbaum Associates, 1982.

Cooper, Charles Henry. *Annals of Cambridge*. 5 vols. Cambridge, 1842–1853.

Douglas, Janet Mary. *The Life and Selections from the Correspondence of William Whewell*. London, 1881.

Farrar, Frederick W. *An Essay on the Origin of Language: Based on Modern Researches, and Especially on the Works of M. Renan*. London, 1860.

Forbes, James David. *The Danger of Superficial Knowledge: An Introductory Lecture to the Course of Natural Philosophy in the University of Edinburgh*. London, 1849.

Gladstone, William E. *Speech on the Commission of Inquiry into the State of the Universities of Oxford and Cambridge*. Oxford, 1850.

Gloucester, Duke of. *Hansard* 22 (April 21, 1834): 979–1009.

Gunning, Henry. *Reminiscences of the University, Town, and County of Cambridge: From the Year 1780*. 2nd ed. 2 vols. London, [1839] 1855.

Hamilton, William. *On the Study of Mathematics*. 2nd ed. London, 1854.

Hare, Augustus J. C. *Memorials of a Quiet Life*. London, 1872.

Hare, Julius Charles. "The Children of Light: A Sermon." Cambridge, 1828. In Hare, *Sermons*, 3–39.

Hare, Julius Charles. *A Vindication of Niebuhr's History of Rome from the Charges of the Quarterly Review.* Cambridge, 1829.
Hare, Julius Charles. "Sermon II: Faith, A Practical Principle." 1840. In Hare, *The Victory of Faith*, 8–15.
Hare, Julius Charles. "Sermon III: Office and Province of Faith." 1840. In Hare, *The Victory of Faith*, 15–25.
Hare, Julius Charles. "Sermon IV: Power of Faith in Man's Natural Life." 1840. In Hare, *The Victory of Faith*, 25–36.
Hare, Julius Charles. "Sermon VI: Faith in Christ, The Victory That Overcometh the World." 1840. In Hare, *The Victory of Faith*, 45–57.
Hare, Julius Charles. *The Victory of Faith and Other Sermons.* Cambridge, 1840. Reprint, Memphis: General Books, 2012.
Hare, Julius Charles. *Miscellaneous Pamphlets on Some of the Leading Questions Agitated in the Church during the Last Ten Years.* Cambridge, 1855.
Hare, Julius Charles. *Charges to the Clergy of the Archdeaconry of Lewes, delivered at the Ordinary Visitations in the Years 1843, 1845 and 1846 with an anonymous Introduction.* Cambridge, 1856.
Hare, Julius Charles. *Sermons Preacht on Particular Occasions.* Cambridge, 1858.
Hare, Julius, and Augustus Hare. *Guesses at Truth.* 2 vols. London, 1827.
Hare-Naylor, Francis. *History of Helvetic Republics.* 2 vols. London, 1801. 2nd enlarged ed. 4 vols. London, 1809.
Hare-Naylor, Francis. *Civil and Military History of Germany, from the Landing of Gustavus to the Treaty of Westphalia.* 2 vols. London, 1816.
Herschel, John. "Mathematics." *Encylopaedia Metropolitana* (1830). Reprinted in *Aspects of the Life and Thought of John Herschel*, edited by Silvian S. Schweber, 434–59. New York: Arno Press, 1989.
Herschel, Margaret. *Lady Herschel: Letters from the Cape, 1834–1838.* Edited by Brian Warner. Cape Town: Friends of the South African Library, 1991.
Hume, David. "Of Miracles" (1748). In *Essays, Literary, Moral and Political by David Hume*, 553–68. London, 1875.
Jones, Richard. *Peasant Rents: Being the First Half of an Essay on the Distribution of Wealth and on Sources of Taxation.* London, 1831.
[Jones, Richard]. "On Tithes." In *The British Magazine*, 2:252–58. London, 1832.
Jones, Richard. *Literary Remains, Consisting of Lectures and Tracts on Political Economy of the Late Rev. Richard Jones.* Edited by John Cazenove, with an introduction by William Whewell. London, 1859.
Lyell, Charles. *Life, Letters, and Journals of Sir Charles Lyell.* Edited by K. Lyell. 2 vols. London, 1881.
Mackintosh, James. *Dissertation Second: Exhibiting a General View of the Progress of Ethical Philosophy with a Preface by William Whewell.* Cambridge, 1835.

Marsh, Herbert. *An Inquiry and an Address to the Members of the Senate of the University of Cambridge, occasioned by the Proposal to Introduce in this Place an Auxiliary Bible Society.* Cambridge, 1811.

Marsh, Herbert. *Consequences of Neglecting to Give the Prayer Book with the Bible.* Cambridge, 1812.

Marx, Karl. *Capital: A Critique of Political Economy.* Vol. 1. Translated by Ben Fowkes. St. Ives, UK: Penguin, [1867] 1990.

Marx, Karl. *Capital: A Critique of Political Economy.* Vol. 2. Translated by David Fernbach. St. Ives, UK: Penguin, [1884] 1992.

Marx, Karl. *Capital: A Critique of Political Economy.* Vol. 3. Translated by David Fernbach. St. Ives, UK: Penguin, [1894] 1991.

Maurice, Frederick, ed. *The Life of Frederick Denison Maurice: Chiefly Told in His Own Letters.* 2 vols. London, 1884.

[McCulloch, John Ramsay]. "Review of Richard Jones, *An Essay on the Distribution of Wealth, and on the Sources of Taxation* (London, 1831)." *Edinburgh Review*, August–December 1831, 84–99.

Merivale, Charles. *Autobiography and Letters of Charles Merivale, Dean of Ely.* Edited by J. A. Merivale. Oxford, 1898.

Michaelis, Johann David. *Introduction to the New Testament, Tr., and Augmented With Notes (And a Dissertation On the Origin and Composition of the Three First Gospels) by Herbert Marsh.* 4 vols. Cambridge, 1801.

Milner, Isaac. *Strictures on Some of the Publications of the rev. Herbert Marsh, Intended as a Reply to His Objections against the British and Foreign Bible Society.* London, 1813.

Milner, Mary. *The Life of Isaac Milner.* London, 1848.

Paine, Thomas. *The Age of Reason*, Pt. 1 (1794). In *The Selected Work of Tom Paine.* Edited by Howard Fast, 281–330. London: Bodley Head, 1948.

Paley, William. *Natural Theology or Evidence of the Existence and Attributes of the Deity, collected from the Appearances of Nature.* 1802. Reprint, Oxford: Oxford University Press, 2006.

Parker, John W., ed. *Essays and Reviews.* London, 1860.

Peacock, D. M. *A Comparative View of the Principles of the Fluxional and Differential Calculus Addressed to the University of Cambridge.* Cambridge, 1819.

Politics of the People: Numbers 1–17. 1848. Reprint, New York: Augustus M. Kelley, 1971.

Rigaud, Stephen Jordan. *A Defence of Halley.* London, 1844.

Rigaud, Stephen Jordan. *Memoir of Stephen P. Rigaud.* Oxford, 1883.

Ritchie, David G. *The Principles of State Interference: Four Essays on the Political Philosophy of Mr. Herbert Spencer, J. S. Mill and T. H. Green.* London, 1891.

Rose, Hugh James. *The State of the Protestant Religion in Germany: In a Series of Discourses Preached before the University of Cambridge.* Cambridge, 1825.

Rose, Hugh James. *Commission and Consequent Duties of the Clergy in a Series of*

Discourses Preached Before the University of Cambridge in April 1828. Cambridge, 1828.

Rose, Hugh James. *The Farmers and the Clergy: Six Letters to the Farmers of England, on Tithes and Church Property.* London, 1831.

Rose, Hugh James. *The Duty of Maintaining the Truth.* Cambridge, 1834.

Schleiermacher, Frederich. *A Critical Essay on the Gospel of St. Luke. By Dr. Frederich Schleiermacher, With an Introduction by the Translator* [Connop Thirlwall]. *Containing an Account of the Controversy respecting the Origin of the Three First Gospels since Bishop Marsh's Dissertation.* London, 1825

Sedgwick, Adam. *A Discourse on the Studies of a University Education.* Cambridge, 1832. Reprint, London: Leicester University Press, 1969.

Sidgwick, Henry. "Philosophy at Cambridge." *Mind* 1 (1876): 235–46.

Sterling, John. *Essays and Tales, by John Sterling. Collected and Edited, with a Memoir of His Life by Julius Charles Hare.* 2 vols. London, 1848.

Sylvester, James J. "Presidential Address: Mathematics and Physics Section." *Report of the Thirty-Ninth Meeting of the BAAS held at Exeter in August 1869.* London, 1870.

Thirlwall, Connop. *History of Greece, from the Earliest Times to the Destruction of Corinth, B.C. 146.* Edited by Peter P. Liddel. 1835. Reprint, Liverpool: Bristol Phoenix Press, 2007.

Thirlwall, Connop. "Works which Remain, and Works which Follow Us." Sermon given at Charterhouse School on Founders Day, December 12, 1845. In Thirlwall, *Remains Literary and Theological,* 3:315–27.

Thirlwall, Connop. *Remains Literary and Theological.* Edited by J. J. Stewart Perowne. 3 vols. London, 1878.

Thirlwall, Connop. *Essays, Speeches and Sermons.* Edited by J. J. Stewart Perowne. London, 1880. This is a reprint of volume 3 from *Remains Literary and Theological.*

Thirlwall, Connop. "The Excellence of Wisdom." Sermon given at Llandingad Church, December 13, 1849. In Thirlwall, *Essays, Speeches and Sermons of Thirlwall,* 340–53.

Thirlwall, Connop. "The Spirit of Truth, The Holy Spirit." Sermon given at University of Cambridge, May 16, 1869. In Thirlwall, *Essays, Speeches and Sermons of Thirlwall,* 387–400.

Thirlwall, Connop. *Letters Literary and Theological of Connop Thirlwall.* Edited by J. J. Stewart Perowne and Louis Stokes. London, 1881.

Todhunter, Issac. *William Whewell, D. D., Master of Trinity College, Cambridge: An Account of His Writings with selections from His Literary and Scientific Correspondence.* 2 vols. London, 1876.

Toynbee, Arnold. "The Ideal Relation of Church and State." Notes of an address delivered at a private meeting in Balliol College, Spring 1879. In Toynbee, *Lectures on the Industrial Revolution,* 249–58.

Toynbee, Arnold. *Lectures on the Industrial Revolution in England, Popular Addresses, Notes, and Other Fragments*. 1884. Reprint, London: Longmans, Green, 1908.

Whately, Richard. *Introduction to Political Economy*. London, 1832.

Whewell, William. *An Elementary Treatise on Mechanics*. Cambridge, 1819.

Whewell, William. "Mathematical Exposition of Some Doctrines of Political Economy." *Transactions Cambridge Philosophical Society* 3 (1830): 191–230.

Whewell, William. *On German Churches. A New Edition. To which is Now Added Notes Written during an Architectural Tour of Picardy and Normandy*. Cambridge, [1830] 1835.

Whewell, William. "Modern Science: Inductive Philosophy." *Quarterly Review* 45 (1831): 374-407.

Whewell, William. *Astronomy and General Physics Considered with Reference to Natural Theology*. London, 1833.

Whewell, William. "Mathematical Exposition of Some Leading Doctrines in Mr Ricardo's Principles of Political Economy." *Transactions Cambridge Philosophical Society* 4 (1833): 155–98.

Whewell, William. *Additional Remarks on Some Parts of Mr Thirlwall's Two Letters on the Admission of Dissenters to Academical Degrees*. Cambridge, 1834.

Whewell, William. *On the Admission of Dissenters to Academical Degrees*. Cambridge, 1834.

Whewell, William. *Thoughts on the Study of Mathematics, as a Part of a Liberal Education*. London, 1835.

Whewell, William. *Newton and Flamsteed: Remarks on an Article in Number CIX of the Quarterly Review*. Cambridge, 1836.

Whewell, William. *History of the Inductive Sciences*. 3 vols. 3rd ed. London, [1837] 1857.

Whewell, William. "Letter to Charles Babbage." *Athenaeum*, May 30, 1837, 1–7.

Whewell, William. *On the Foundations of Morals: Four Sermons Preached before the University of Cambridge, November 1837*. New York, 1839.

Whewell, William. *Aphorisms Concerning Ideas, Science, and the Language of Science*. London, 1840.

Whewell, William. *The Philosophy of the Inductive Sciences Founded Upon their History*. 2 vols. Cambridge, 1840.

Whewell, William. *The Elements of Morality, Including Polity*. 3rd ed. Cambridge, [1841] 1854.

Whewell, William. "Lecture 1: Definition of a System of Morality." In Whewell, *Lectures on Systematic Morality*.

Whewell, William. "Lecture VI: Virtue, Happiness, Conscience and the Will of God." In Whewell, *Lectures on Systematic Morality*, 38–54.

Whewell, William. "Lecture VIII: The Relation of Church and State." In Whewell, *Lectures on Systematic Morality*, 54–61.

Whewell, William. *Lectures on Systematic Morality*. London, 1846. Reprint, Memphis, TN: General Books, 2012.

Whewell, William. *The Christian's Duty Towards Transgressors: Sermon Preached in the Chapel of the Philanthropic Society, St. Georges Fields, on Sunday, May 16, 1847.* London, 1847.

Whewell, William. *Of a Liberal Education in General, and with Particular Reference to the Leading Studies of the University of Cambridge*. Cambridge, 1848.

Whewell, William. "On Hegel's Criticism of Newton's Principia." *Transactions of the Cambridge Philosophical Society* 51 (1849): 696–706.

Whewell, William. Review of Babington Macaulay, "In Praise of Superficial Knowledge" (1846). *Fraser's Magazine*, August 1849, 171–75.

Whewell, William. *Of the Plurality of Worlds*. 4th ed. London, [1853] 1855.

Whewell, William. "On the Influence of the History of Science upon Intellectual Education." In *Lectures on Education, delivered at the Royal Institution of Great Britain*, 3–36. London, 1854.

Whewell, William. *On the Material Aids of Education: Being an Inaugural Lecture on the Occasion of the Educational Exhibition of 1854*. London, 1854.

Whewell, William. *The Platonic Dialogues for English Readers*. Cambridge, 1859.

Whewell, William, ed. *The Mathematical Works of Isaac Barrow*. Cambridge, 1860.

Whewell, William. *On the Philosophy of Discovery: Chapters Historical and Critical*. London, 1860.

Whewell, William. *Six Lectures on Political Economy*. Cambridge, 1862.

Whewell, William. "Of Certain Analogies between Architecture and the Other Fine Arts." *Royal Institute of British Architects*, session 1862–1863, 175–81.

Whewell, William. "Comte and Positivism." *Macmillan's Magazine*, 1866, 353–62.

Wordsworth, Christopher. "Diary of Christopher Wordsworth, 1819–1841." TC: 0.12.55.

SECONDARY SOURCES

Aarsleff, Hans. *The Study of Language in England, 1780–1860*. Minneapolis: University of Minnesota Press, [1967] 1983.

Abrams, M. H. *The Mirror and the Lamp: Romantic Theory and the Critical Tradition*. Oxford: Oxford University Press, [1953] 1971.

Ackerberg-Hastings, Amy. "Analysis and Synthesis in John Playfair's 'Elements of Geometry.'" *British Journal for the History of Science* 35, no. 1 (2002): 43–72.

Adorno, Theodor W., and Max Horkheimer. *Dialectic of Enlightenment*. London: Verso, [1944] 2008.

Agar, Jon. *The Government Machine: A Revolutionary History of the Computer.* Cambridge, MA: MIT Press, 2003.

Alborn, Timothy L. "The Business of Induction: Industry and Genius in the Language of British Scientific Reform, 1820–1840." *History of Science* 34, no. 1 (March 1996): 91–121.

Allen, Peter. *The Cambridge Apostles: The Early Years.* Cambridge: Cambridge University Press, 2010.

Altholz, Josef L. "The Mind of Victorian Orthodoxy: Anglican Responses to 'Essays and Reviews,' 1860–1864." *Church History* 51, no. 2 (June 1982): 186–97.

Altholz, Josef L. *Anatomy of a Controversy: The Debate over "Essays & Reviews."* Aldershot, UK: Scolar Press, 1994.

Altholz, Josef L. "A Tale of Two Controversies: Darwinism in the Debate over 'Essays and Reviews.'" *Church History* 63, no. 1 (March 1994): 50–59.

Appleby, Joyce. *The Relentless Revolution: A History of Capitalism.* New York: W. W. Norton, 2010.

Armytage, W. H. G. *Four Hundred Years of English Education.* 2nd ed. Cambridge: Cambridge University Press, [1964] 1970.

Ashby, Eric, and Mary Anderson. "Introduction." In Sedgwick, *A Discourse on the Studies of a University Education*, 7–25.

Ashworth, William J. "The Calculating Eye: Baily, Herschel, Babbage and the Business of Astronomy." *British Journal for the History of Science* 27, no. 4 (1994): 409–41.

Ashworth, William J. "Memory, Efficiency, and Symbolic Analysis: Charles Babbage, John Herschel, and the Industrial Mind." *Isis* 87, no. 4 (1996): 629–53.

Ashworth, William J. "'Labour Harder than *Thrashing*': John Flamsteed, Property and Intellectual Labour in Nineteenth-Century England." In Willmoth, *Flamsteed's Stars*, 199–216.

Ashworth, William J. "Analytical Society (Act. 1812–1813)." *Oxford Dictionary of National Biography.*

Attis, David. *Mathematics and the Making of Modern Ireland: Trinity College Dublin from Cromwell to the Celtic Tiger.* Boston: Docent Press, 2014.

Augstein, H. F. "Prichard, James Cowles (1786–1848)." *Oxford Dictionary of National Biography.*

Bagwell, Philip S. *The Transport Revolution from 1770.* London: B. T. Batsford, 1974.

Banham, Gary. *Kant's Transcendental Imagination.* Basingstoke, UK: Palgrave Macmillan, 2005.

Becher, Harvey W. "William Whewell and Cambridge Mathematics." *Historical Studies in the Physical Sciences* 11, no. 1 (January 1980): 1–48.

Becher, Harvey W. "William Whewell's Odyssey: From Mathematics to Moral Philosophy." In Fisch and Shaffer, *Whewell: A Composite Portrait*, 1–30.

Becher, Harvey W. "Radicals, Whigs, and Conservatives: The Middle and Lower Classes in the Analytical Revolution in Cambridge in the Age of Aristocracy." *British Journal for the History of Science* 28, no. 4 (December 1995): 405–26.

Beck, Naomi. *Hayek and the Evolution of Capitalism*. Chicago: University of Chicago Press, 2018.

Behler, Ernst. *Fichte, Jacobi, and Schelling: Philosophy of German Idealism*. New York: Continuum, 1987.

Beiser, Frederick C. *German Idealism: The Struggle against Subjectivism 1781–1801*. Cambridge, MA: Harvard University Press, 2002.

Belchem, John. "Henry Hunt and the Evolution of the Mass Platform." *English Historical Review* 93, no. 369 (October 1978): 739–73.

Belchem, John. *"Orator" Hunt: Henry Hunt and English Working Class Radicalism*. London: Breviary Stuff Publications, 2012.

Bloor, David. "Hamilton and Peacock on the Essence of Algebra." In Mehrtens, Bos, and Schneider, *Social History of Nineteenth-Century Mathematics*, 202–31.

Brent, Richard. *Liberal Anglican Politics: Whiggery, Religion, and Reform 1830–1841*. Oxford: Oxford University Press, 1987.

Bridenthal, Renate. "Was There a Roman Homer? Niebuhr's Thesis and Its Critics." *History and Theory* 11, no. 2 (1972): 193–213.

Briefs, Goetz A. "The Economic Philosophy of Romanticism." *Journal of the History of Ideas* 2, no. 3 (June 1941): 279–300.

Brock, W. H., and R. M. Macleod. "The Scientists' Declaration: Reflexions on Science and Belief in the Wake of 'Essays and Reviews,' 1864–5." *British Society for the History of Science* 9, no. 1 (1976): 39–66.

Brooke, John Hedley. *Science and Religion: Some Historical Perspectives*. Cambridge: Cambridge University Press, 1991.

Brose, Oliver. "F. D. Maurice and the Victorian Crisis of Belief." *Victorian Studies* 3, no. 3 (March 1960): 227–48.

Brown, Wendy. *Undoing the Demos: Neoliberalism's Stealth Revolution*. New York: Zone Books, 2015.

Buchanan, Alexandrina. *Robert Willis and the Foundation of Architectural History*. Cambridge: Cambridge University Press, 2013.

Burbridge, David. "William Paley Confronts Erasmus Darwin: Natural Theology and Evolutionism in the Eighteenth Century." *Science and Christian Belief* 10, no. 1 (April 998): 49–71.

Burrow, J. W. "The Uses of Philology in Victorian England." In Robson and Cannon, *Ideas and Institutions*, 180–204.

Button, Graham, Jeff Coulter, John R. E. Lee, and Wes Sharrock. *Computers, Minds and Conduct.* Cambridge: Polity Press, 1995.

Caldwell, Bruce. "F. A. Hayek and the Economic Calculus." *History of Political Economy* 48, no. 1 (2016): 151–80.

Cannon, Susan Faye. *Science in Culture: The Early Victorian Period.* New York: Dawson, 1978.

Cannon, Walter F. "John Herschel and the Idea of Science." *Journal of the History of Ideas* 22 (April–June 1961): 215–39.

Cannon, Walter F. "History in Depth: The Early Victorian Period." *History of Science* 3, no. 1 (1964): 20–38.

Cannon, Walter F. "Scientists and Broad Churchmen: An Early Victorian Intellectual Network." *Journal of British Studies* 4, no. 1 (1964): 65–88.

Cantor, Geoffrey N. "Between Rationalism and Romanticism: Whewell's Historiography of the Inductive Sciences." In Fisch and Schaffer, *William Whewell,* 67–86.

Carlyle, Thomas. *The Life of John Sterling.* London, 1851.

Carter, Matt. *T. H. Green and the Development of Ethical Socialism.* Exeter, UK: Imprint Academic, 2003.

Chadwick, Henry. "Romanticism and Religion." In Chadwick, *Tradition and Exploration,* 217–28.

Chadwick, Henry. *Tradition and Exploration: Collected Papers on Theology and the Church.* Norwich, UK: Canterbury Press, 1994.

Chadwick, Owen. *The Victorian Church: Part One 1829–1859.* 3rd ed. 1966. Reprint, London: SCM Press, 1971.

Chase, Malcolm. *The People's Farm: English Radical Agrarianism 1775–1840.* Oxford: Oxford University Press, 1988.

Christensen, Torben. *Origin and History of Christian Socialism 1848–54.* Aarhus, Denmark: Universitetsforlaget I Aarhus Press, 1962.

Clark, G. Kitson. *The Making of Victorian England.* London: Routledge, 1962.

Clark, J. C. D. *English Society 1688–1832.* Cambridge: Cambridge University Press, 1985.

Clark, J. W., and H. C. G. Matthew. "Thirlwall, (Newell) Connop (1797–1875)." *Oxford Dictionary of National Biography.*

Collini, Steffani. *Matthew Arnold: A Critical Portrait.* Oxford: Oxford University Press, [1988] 1994.

Collins, Francis. *The Language of God: A Scientist Presents Evidence for Belief.* London: Simon and Schuster, 2007.

Collins, Harry. *Artificial Intelligence: Against Humanity's Surrender to Computers.* Cambridge: Polity Press, 2018.

Corsi, Pietro. *Science and Religion: Baden Powell and the Anglican Debate, 1800–1860.* Cambridge: Cambridge University Press, 1988.

Cranston, Maurice. *The Romantic Movement.* Oxford: Oxford University Press, 1994.

Crowe, Michael J. *The Extraterrestrial Life Debate 1750–1900: The Idea of a Plurality of Worlds from Kant to Lowell.* Cambridge: Cambridge University Press, 1988.

Crowe, Michael J. "William Whewell, the Plurality of Worlds, and the Modern Solar System." *Journal of Religion and Society* 51, no. 2 (2016): 431–49.

Cunningham, Andrew, and Nicholas Jardine, eds. *Romanticism and the Sciences.* Cambridge: Cambridge University Press, 1990.

Curran, Stuart, ed. *The Cambridge Companion to British Romanticism.* Cambridge: Cambridge University Press, 1993.

Davies, William. *The Limits of Neoliberalism: Authority, Sovereignty, and the Logic of Competition.* London: Sage, 2014.

Davies, William. *Nervous States: How Feeling Took over the World.* London: Jonathan Cape, 2018.

Dawkins, Richard. *The Selfish Gene.* Oxford: Oxford University Press, 1976.

Dawkins, Richard. *The God Delusion.* 2006. Reprint, Suffolk, UK: Transworld, 2016.

Dawson, P. M. S. "Poetry in an Age of Revolution." In Curran, *The Cambridge Companion to British Romanticism*, 48–73.

Deacon, Richard. *The Cambridge Apostles: A History of Cambridge University's Elite Intellectual Secret Society.* London: R. Royce, 1985.

Dembski, William A. *Intelligent Design: The Bridge between Science and Theology.* Downers Grove, IL: InterVarsity Press, 1999.

Desmond, Adrian. *Archetypes and Ancestors: Palaeontology in Victorian London 1850–1875.* London: Blond and Briggs, 1982.

Desmond, Adrian. *The Politics of Evolution: Morphology, Medicine, and Reform in Radical.* Chicago: University of Chicago Press, 1989.

Desmond, Adrian. *Huxley: The Devil's Disciple.* London: Michael Joseph, 1994.

Desmond, Adrian, and James Moore. *Darwin.* London: Penguin, 1991.

Desmond, Adrian, and James Moore. *Darwin's Sacred Cause: Race, Slavery and the Quest for Human Origins.* London: Allen Lane, 2009.

Distad, N. Merrill. "The Philological Museum of 1831–1833." *Victorian Periodicals Newsletter* 5 (December 1972): 27–30.

Distad, N. Merrill. *Guessing at Truth: The Life of Julius Hare 1795–1855.* Shepherdstown, WV: Patmos Press, 1979.

Distad, N. Merrill. "Hare, Julius Charles (1795–1855)." *Oxford Dictionary of National Biography.*

Doyle, William. *The Oxford History of the French Revolution.* 3rd ed. Oxford: Oxford University Press, [1989] 2018.

Dubbey, J. M. *The Mathematical Work of Charles Babbage*. Cambridge: Cambridge University Press, 1978.

Eddy, Matthew D., and David Knight. "Introduction." In Paley, *Natural Theology*, ix–xxxv.

Enros, Phillip. "The Analytical Society: Mathematics at Cambridge University in the Early Nineteenth Century." PhD diss., University of Toronto, 1979.

Enros, Phillip. "The Analytical Society (1812–1813): Precursor of the Renewal of Cambridge Mathematics." *Historia Mathematica* 10, no. 1 (February 1983): 24–47.

Evra, James Van. "Richard Whately and Logical Theory." In Gabbay and Woods, *Handbook of the History of Logic*, 75–91.

Faber, Geoffrey. *Oxford Apostles: A Character Study of the Oxford Movement*. 1933. Reprint, London: Faber and Faber, 1974.

Fairchild, Hoxie N. "Romanticism and the Religious Revival in England." *Journal of the History of Ideas* 2, no. 3 (June 1941): 330–38.

Fisch, Menachem. "A Philosopher's Coming of Age: A Study in Esoteric Intellectual History." In Fisch and Schaffer, *William Whewell*, 31–66.

Fisch, Menachem. "The Problematic History of Nineteenth-Century British Algebra." *British Journal for the History of Science* 27, no. 3 (September 1994): 247–76.

Fisch, Menachem, and Simon Schaffer, eds. *William Whewell: A Composite Portrait*. Oxford: Oxford University Press, 1991.

Forbes, Duncan. *The Liberal Anglican Idea of History*. Cambridge: Cambridge University Press, [1952] 2006.

Forrest, Robert K. "Marsh, Herbert (1757–1839)." *Oxford Dictionary of National Biography*.

Foster, John Bellamy, Brett Clark, and Richard York. *Critique of Intelligent Design: Materialism versus Creationism from Antiquity to the Present*. New York: Monthly Review Press, 2008.

Franchi, Stefano, and Francesco Bianchini, eds. *The Search for a Theory of Cognition in Early Mechanisations and New Ideas*. Amsterdam: Rodopi, 2011.

Franchi, Stefano, and Güven Güzeldere, eds. *Mechanical Bodies, Computational Minds: Artificial Intelligence from Automata to Cyborgs*. Cambridge, MA: MIT Press, 2005.

Francis, Mark. "The Origins of Essays and Reviews: An Interpretation of Mark Pattison in the 1850s." *Historical Journal* 17, no. 4 (December 1974): 797–811.

Franklin, Michael J. *'Orientalist Jones': Sir William Jones, Poet, Lawyer, and Linguist, 1746–1794*. Oxford: Oxford University Press, 2011.

Fyfe, Aileen. "The Reception of William Paley's *Natural Theology* in the University of Cambridge." *British Journal for the History of Science* 30, no. 3 (September 1997): 321–35.

Gabbay, Dov M., and John Woods, eds. *Handbook of the History of Logic*. Vol. 4, *British Logic in the Nineteenth Century*. Amsterdam: Elsevier, 2008.

Garnett, Richard. "Donaldson, John William (1811–1861)." Revised by Richard Smail. *Oxford Dictionary of National Biography*.

Garrard, J. R. "Thorp, Thomas (1797–1877)." *Oxford Dictionary of National Biography*.

Gascoigne, John. "Mathematics and Meritocracy: The Emergence of the Cambridge Mathematical Tripos." *Social Studies of Science* 14, no. 4 (November 1984): 547–84.

Gascoigne, John. *Cambridge in the Age of the Enlightenment: Science, Religion and Politics from the Restoration to the French Revolution*. Cambridge: Cambridge University Press, 1989.

Gibson, William. *Church, State and Society, 1760–1850*. Basingstoke, UK: Palgrave Macmillan, 1994.

Gibson, William. *The Church of England 1688–1832: Unity and Accord*. London: Routledge, 2001.

Gillespie, Neal C. "Divine Design and the Industrial Revolution: William Paley's Abortive Reform of Natural Theology." *Isis* 81, no. 2 (June 1990): 214–29.

Giustino, David de. "Finding an Archbishop: The Whigs and Richard Whately in 1831." *Church History* 64, no. 2 (June 1995): 218–36.

Gould, Stephen Jay. *Rocks of Ages: Science and Religion in the Fullness of Life*. 1999. Reprint, London: Ballantine Books, 2001.

Grattan-Guiness, Ivor. "Charles Babbage: Production by Numbers." *Annals of Science* 47, no. 1 (1990): 81–87.

Gray, Sara. *The Dictionary of British Women Artists*. Cambridge: Lutterworth Press, 2009.

Green, Francis, and David Kynaston. *Engines of Privilege: Britain's Private School Problem*. London: Bloomsbury, 2019.

Gregory, Frederick. "Theology and the Sciences." In Cunningham and Jardine, *Romanticism and the Sciences*, 69–81.

Grier, David Alan. *When Computers Were Human*. Princeton, NJ: Princeton University Press, 2005.

Grossman, Henryk. "The Evolutionist Revolt against Classical Economics: II. In England–James Steuart, Richard Jones, Karl Marx." *Journal of Political Economy* 51, no. 6 (December 1943): 506–33.

Guicciardini, Niccolo. *Reading the Principia: The Debate of Newton's Mathematical Methods for Natural Philosophy from 1687–1736*. 1999. Reprint, Cambridge: Cambridge University Press, 2003.

Guiney, Louise Imogen. *Hurrell Froude: Memoranda and Comments*. London: Methuen, 1904.

Hall, Michael. "What Do Victorian Churches Mean? Symbolism and Sacramentalism in Anglican Church Architecture, 1850–1870." *Journal of the Society of Architectural Historians* 59, no. 1 (2000): 78–95.

Hamlin, Christopher. *Public Health and Social Justice in the Age of Chadwick: Britain, 1800–1854.* Cambridge: Cambridge University Press, 1998.

Harrison, Peter. *The Fall of Man and the Foundations of Science.* Cambridge: Cambridge University Press, 2007.

Harrison, Peter. *The Territories of Science and Religion.* Chicago: University of Chicago Press, 2015.

Havens, Michael K. "Coleridge on the Evolution of Language." *Studies in Romanticism* 20, no. 4 (Winter 1981): 163–83.

Hayek, Friedrich. *The Sensory Order: An Inquiry into the Foundations of Theoretical Psychology.* Chicago: University of Chicago Press, 1952.

Heathcote, A. W. "William Whewell's Philosophy of Science." *British Journal for the Philosophy of Science* 4, no. 16 (February 1954): 302–14.

Hedley, Douglas. *Living Forms of the Imagination.* New York: T and T Clark International, 2008.

Higgitt, Rebekah. "Astronomers against Newton." *Endeavour* 28 (March 2004): 20–24.

Higgitt, Rebekah. *Recreating Newton: Newtonian Biography and the Making of Nineteenth-Century History of Science.* 2007. Reprint, Pittsburgh: University of Pittsburgh Press, 2020.

Hill, Harvey. "Religion and the University: The Controversy over *Essays and Reviews* at Oxford." *Journal of the American Academy of Religion* 73, no. 1 (March 2005): 183–207.

Hilton, Boyd. *The Age of Atonement: The Influence of Evangelicalism on Social and Economic Thought 1785–1865.* Oxford: Oxford University Press, 1988.

Hilton, Boyd. "Whiggery, Religion and Social Reform: The Case of Lord Morpeth." *Historical Journal* 37, no. 4 (December 1994): 829–59.

Hilton, Boyd. *A Mad, Bad & Dangerous People? England 1783–1846.* Oxford: Oxford University Press, 2008.

Hirschman, Albert O. *The Passions and the Interests: Political Arguments for Capitalism before Its Triumph.* 1977. Reprint, Princeton, NJ: Princeton University Press, 2013.

Hirschman, Albert O. *The Rhetoric of Reaction: Perversity, Futility and Jeopardy.* Cambridge, MA: Belknap Press of Harvard University Press, 1991.

Hole, Robert. *Pulpits, Politics and Public Order in England 1760–1832.* Cambridge: Cambridge University Press, 1989.

Hopkins, Hugh Evans. *Charles Simeon of Cambridge.* London: Hodder and Stoughton, 1977.

Hoskins, Michael. "Astronomers at War: South versus Sheepshanks." *Journal of the History of Astronomy* 20, no. 3 (1989): 175–212.

Iliffe, Rob. *Priest of Nature: The Religious Worlds of Isaac Newton*. Oxford: Oxford University Press, 2017.

Jackson, Ben. *Equality and the British Left: A Study in Progressive Political Thought, 1900–64*. Manchester: Manchester University Press, 2007.

Jackson, Ben. "Currents of Neo-Liberalism: British Political Ideologies and the New Right, c. 1955–1979." *English Historical Review* 131, no. 551 (August 2016): 823–50.

Jackson, H. J. "Etymology and Etymologic." *Journal of the History of Ideas* 44, no. 1 (1983): 75–88.

Jacquette, Dale. "Boole's Logic." In Gabbay and Woods, *Handbook of the History of Logic*, 331–79.

Jessop, Ralph. "The Logic of Sir William Hamilton: Tunnelling through Sand to Place the Keystone in the Aristotelic Arch." In Gabbay and Woods, *Handbook of the History of Logic*, 93–162.

Johnson, Phillip E. *The Wedge of Truth: Splitting the Foundations of Naturalism*. Downers Grove, IL: InterVarsity Press, 2000.

Jones, Daniel Stedman. *Masters of the Universe: Hayek, Friedman, and the Birth of Neoliberal Politics*. Princeton, NJ: Princeton University Press, 2012.

Jones, Gareth Stedman. *An End to Poverty?* London: Profile Books, 2004.

Jong, Willem R. de. "Kant's Theory of Geometrical Reasoning and the Analytic-Synthetic Distinction: On Hintikka's Interpretation of Kant's Philosophy of Mathematics." *Studies in History and Philosophy of Science* 28, no. 1 (1997): 141–66.

Jordan, Alexander. "Thomas Carlyle and Political Economy: The 'Dismal Science' in Context." *English Historical Review* 132, no. 555 (April 2017): 286–317.

Joyce, Patrick. *The State of Freedom: A Social History of the British State*. Cambridge: Cambridge University Press, 2013.

Katz, Solomon. "Remembrance of Things Past." *Pacific Historical Review* 38, no. 1 (February 1969): 1–20.

Keach, William. "Romanticism and Language." In Curran, *The Cambridge Companion to British Romanticism*, 95–119.

Kirby, James. "R. H. Tawney and Christian Social Teaching: Religion and the Rise of Capitalism Reconsidered." *English Historical Review* 131, no. 551 (August 2016): 793–822.

Knox, Kevin. "Dephlogisticating the Bible: Natural Philosophy and Religious Controversy in Late Georgian Cambridge." *History of Science* 34 (1996): 167–200.

Knox, Kevin. "The Negative Side of Nothing: Edward Waring, Isaac Milner and Newtonian Values." In Knox and Noakes, *From Newton to Hawking*, 205–40.

Knox, Kevin, and Richard Noakes, eds. *From Newton to Hawking: A History of Cambridge University's Lucasian Professors of Mathematics*. Cambridge: Cambridge University Press, 2003.

Koot, Gerard M. *English Historical Economists, 1870–1926: The Rise of Economic History and Neomercantilism*. Cambridge: Cambridge University Press, 1987.

Koot, Gerard M. "Cunningham, William (1849–1919)." *Oxford Dictionary of National Biography*.

Kooy, Michael John. "Romanticism and Coleridge's Idea of History." *Journal of the History of Ideas* 60, no. 4 (October 1999): 717–35.

Kubo, Shin. "George Pryme, Dugald Stewart, and Political Economy at Cambridge." *History of Political Economy* 45, no. 1 (2013): 61–97.

Kuhn, Thomas. *The Structure of Scientific Revolutions*. Chicago: University of Chicago Press, 1962.

Lambert, Kevin. "A Natural History of Mathematics: George Peacock and the Making of English Algebra." *Isis* 104, no. 2 (2013): 278–302.

Leask, Nigel. *The Politics of Imagination in Coleridge's Critical Thought*. Basingstoke, UK: Macmillan Press, 1988.

Lee, Sidney. "Walker, (William) Sidney (1795–1846)." Revised by Penelope Hicks. *Oxford Dictionary of National Biography*.

Lewis, Michael A. "Charles Babbage: Reclaiming an Operations Management Pioneer." *Journal of Operations Management* 25, no. 2 (March 2007): 248–59.

Linker, Damon. "From Kant to Schelling: Counter-Enlightenment in the Name of Reason." *Review of Metaphysics* 54, no. 2 (December 2000): 337–77.

Maas, Harro. "'A Hard Battle to Fight': Natural Theology and the Dismal Science, 1820–50." *History of Political Economy* 40, no. 5 (2008): 143–67.

Macfarlane, Alan. *F.W. Maitland and the Making of the Modern World*. Cambridge: CreateSpace Independent Publishing Platform, 2018.

MacHale, Desmond. *The Life and Work of George Boole: A Prelude to the Digital Age*. 1985. Reprint, Cork: Cork University Press, 2014.

Mackinnon, D. M. "Some Aspects of the Treatment of Christianity by the British Idealists." *Religious Studies* 20, no. 1 (1984): 133–44.

Mayr, Otto. *Authority, Liberty and Automatic Machinery in Early Modern Europe*. Baltimore: John Hopkins University Press, 1986.

McCloskey, Deirdre N. *Bourgeois Dignity: Why Economics Can't Explain the Modern World*. Chicago: University of Chicago Press, 2010.

McFarland, G. F. "The Early Literary Career of Julius Charles Hare." *Bulletin of the John Rylands Library* 46, no. 1 (September 1963): 42–83.

McFarland, G. F. "Julius Hare: Coleridge, DeQuincey, and German Literature." *Bulletin of the John Rylands Library* 47, no. 1 (1964): 165–97.

Mehrtens, Herbert, Henk Bos, and Ivo Schneider, eds. *Social History of Nineteenth-Century Mathematics*. Boston: DA Information Services, 1981.

Milnes, Tim. "Coleridge's Logic." In Gabbay and Woods, *Handbook of the History of Logic*, 37–74.

Moggach, Douglas, and Gareth Stedman Jones, eds. *The 1848 Revolutions and European Political Thought*. Cambridge: Cambridge University Press, 2020.

Mokyr, Joel. *The Enlightened Economy: Britain and the Industrial Revolution, 1700–1850.* New Haven, CT: Yale University Press, 2011.

Mokyr, Joel. *A Culture of Growth: The Origins of the Modern Economy.* Princeton, NJ: Princeton University Press, 2018.

Moore, D. C. "Social Structure, Political Structure, and Public Opinion in Mid-Victorian England." In Robson and Cannon, *Ideas and Institutions of Victorian Britain,* 20–57.

Morrell, John B. "Theophobia Gallica: Natural Philosophy, Religion, and Politics in Edinburgh, 1789–1815." *Notes and Records of the Royal Society of London* 26, no. 1 (June 1971): 43–63.

Morus, Iwan Rhys. "Manufacturing Nature: Science, Technology and Victorian Consumer Culture." *British Journal for the History of Science* 29, no. 4 (1996): 403–34.

Murphy, G. Martin. "Malkin, Benjamin Heath (1769–1842)." *Oxford Dictionary of National Biography.*

Nockles, Peter B. "Rose, Hugh James (1795–1838)." *Oxford Dictionary of National Biography.*

Olson, Richard. *British Natural Philosophy and British Physics: A Study in the Foundations of the Victorian Style.* Princeton, NJ: Princeton University Press, 1975.

Oslington, Paul. "Natural Theology, Theodicy, and Political Economy in Nineteenth-Century Britain: William Whewell's Struggle." *History of Political Economy* 49, no. 4 (December 2017): 575–606.

Oxford Dictionary of National Biography. www.oxforddnb.com.

Panteki, Maria. "French 'Logique' and British 'Logic': On the Origins of Augustus De Morgan's Early Logical Inquiries, 1805–1835." In Gabbay and Woods, *Handbook of the History of Logic,* 381–456.

Parry, Jonathan. *The Rise and Fall of Liberal Government in Victorian Britain.* New Haven, CT: Yale University Press, 1993.

Pawley, Margaret. "Digby, Kenelm Henry (1795/6–1880)." *Oxford Dictionary of National Biography.*

Peacock, A. J. *Bread or Blood: A Study of the Agrarian Riots in East Anglia in 1816 with a Foreword by E. P. Thompson.* London: V. Gollancz, 1965.

Piketty, Thomas. *Capital in the Twenty-First Century.* Cambridge, MA: Harvard University Press, 2014.

Pinkard, Terry. *German Philosophy 1760–1860: The Legacy of Idealism.* Cambridge: Cambridge University Press, 2002.

Pocock, J. G. A. *Virtue, Commerce, and History: Essays on Political Thought and History, Chiefly in the Eighteenth Century.* Cambridge: Cambridge University Press, 1985.

Porter, Theodore M. *Trust in Numbers: The Pursuit of Objectivity in Science and Public Life.* Princeton, NJ: Princeton University Press, 1996.

Powell, J. P. "Some Nineteenth-Century Views on the University Curriculum." *History of Education Quarterly* 5, no. 2 (June 1965): 97–109.

Preyer, Robert O. "The Romantic Tide Reaches Trinity: Notes on the Transmission and Diffusion of New Approaches to Traditional Studies at Cambridge, 1820–1840." *Annals of the New York Academy of Sciences* 360, no. 1 (April 1981): 39–68.

Preyer, Robert O. "John Stuart Mill on the Utility of Classical Greece." *Browning Institute Studies* 10 (1982): 41–70.

Preyer, Robert O. "Breaking Out: The English Assimilation of Continental Thought in Nineteenth-Century Rome." *Browning Institute Studies* 12 (1984): 53–72.

Preyer, Robert O. "The Language of Discovery: William Whewell and George Eliot." *Browning Institute Studies* 16 (1988): 123–52.

Prickett, Stephen. *Coleridge and Wordsworth: The Poetry of Growth.* Cambridge: Cambridge University Press, 1970.

Prickett, Stephen. *Words and the Word: Language, Poetics and Biblical Interpretation.* Cambridge: Cambridge University Press, 1988.

Pullen, J. M. "Jones, Richard (1790–1855)." *Oxford Dictionary of National Biography.*

Pycior, Helena. "George Peacock and the British Origins of Symbolical Algebra." *Historia Mathematica* 8, no. 1 (February 1981): 23–45.

Pycior, Helena. "Early Criticism of the Symbolical Approach to Algebra." *Historia Mathematica* 9, no. 4 (November 1982): 392–412.

Quinn, Aleta. "William Whewell's Philosophy of Architecture and the Historicization of Biology." *Studies in History and Philosophy of Biological and Biomedical Science* 59 (October 2016): 11–19.

Rajan, Supritha. *A Tale of Two Capitalisms: Sacred Economics in Nineteenth-Century Britain.* Ann Arbor: University of Michigan Press, 2015.

Ramsey, Arthur Michael. *F. D. Maurice and the Conflicts of Modern Theology.* Eugene, OR: WIPF & STOCK, [1951] 2011.

Raub, Cymbre Quincy. "Robert Chambers and William Whewell: A Nineteenth-Century Debate over the Origin of Language." *Journal of the History Ideas* 49, no. 2 (June 1988): 287–300.

Richards, Joan L. *Mathematical Visions: The Pursuit of Geometry in Victorian England.* London: Academic Press, 1988.

Richards, Joan L. "Rigor and Clarity: Foundations of Mathematics in France and England, 1800–1840." *Science in Context* 4, no. 2 (1991): 297–319.

Richards, Robert J. *The Romantic Conception of Life: Science and Philosophy in the Age of Goethe.* Chicago: University of Chicago Press, 2002.

Richeson, A. W. "On Faraday's Terminology in Electrolysis." *Isis* 36, no. 3–4 (1946): 160–62.

Richter, Melvin. *The Politics of Conscience: T. H. Green and His Age*. 1964. Reprint, Bristol: Thoemmes Continuum, 1996.

Ridler, Ann Margaret. "Latham, Robert Gordon (812–1888)." *Oxford Dictionary of National Biography*.

Robson, Robert, and Walter F. Cannon. "William Whewell, FRS (1794–1866)." *Notes & Records of the Royal Society* 19, no. 2 (December 1964): 168–91.

Robson, Robert, and Walter F. Cannon, eds. *Ideas and Institutions of Victorian Britain: Essays in Honour of George Kitson Clark*. London: Bell, 1967.

Robson, Robert, and Walter F. Cannon. "Trinity College in the Age of Peel." In Robson and Cannon, *Ideas and Institutions*, 312–35.

Roe, Nicholas. *Wordsworth and Coleridge: The Radical Years*. Oxford: Oxford University Press, 1988.

Rogan, Tim. *The Moral Economists: R. H. Tawney, Karl Polanyi, E. P. Thompson, and the Critique of Capitalism*. Princeton, NJ: Princeton University Press, 2017.

Rogerson, John. *Old Testament Criticism in the Nineteenth Century: England and Germany*. London: SPCK, 1984.

Rogerson, John. *The Bible and Criticism in Victorian Britain: Profiles of F. D. Maurice and William Robertson Smith*. Sheffield, UK: Sheffield Academic Press, 1995.

Roll, Eric. *A History of Economic Thought*. 5th ed. London: Faber and Faber, [1938] 1992.

Ross, S. "Faraday Consults the Scholars: The Origins of the Terms of Electrochemistry." *Notes and Records of the Royal Society of London* 16, no. 2 (November 1961): 187–220.

Rothblatt, Sheldon. "The Student Sub-culture and the Examination System in Early Nineteenth Century Oxbridge." In Stone, *Oxford and Cambridge*, 247–303.

Rousseau, G. S., ed. *The Languages of Psyche: Mind and Body in Enlightenment Thought*. Berkeley: University of California Press, 1987.

Rudwick, Martin J. S. *Earth's Deep History: How It Was Discovered and Why It Matters*. Chicago: University of Chicago Press, 2014.

Rule, Philip C. "Coleridge's Reputation as a Religious Thinker: 1816–1972." *Harvard Theological Review* 67, no. 3 (July 1974): 289–320.

Ruse, Michael. "Charles Lyell and the Philosophers of Science." *British Journal for the History of Science* 9, no. 4 (December 1976): 121–31.

Sanders, Charles Richard. "Coleridge, Maurice and the Church Universal." *Journal of Religion* 21, no. 1 (January 1941): 31–45.

Sanders, Charles Richard. *Coleridge and the Broad Church Movement*. Durham, NC: Duke University Press, 1942.

Sandys, John Edwin. *A History of Classical Scholarship: The Eighteenth Century in Germany and the Nineteenth Century in Europe and the United States of Amer-*

ica. 3 vols. 1903–1908. Reprint, Cambridge: Cambridge University Press, 2010.

Schaffer, Simon. "States of Mind: Enlightenment and Natural Philosophy." In Rousseau, *The Languages of Psyche*, 223–90.

Schaffer, Simon. "Astronomers Mark Time: Discipline and the Personal Equation." *Science in Context* 2, no. 1 (1988): 115–45.

Schaffer, Simon. "Genius in Romantic Natural Philosophy." In Cunningham and Jardine, *Romanticism and the Sciences*, 82–98.

Schaffer, Simon. "The History and Geography of the Intellectual World: Whewell's Politics of Language." In Fisch and Schaffer, *William Whewell*, 201–31.

Schaffer, Simon. "Babbage's Intelligence: Calculating Engines and the Factory System." *Critical Inquiry* 21, no. 1 (Autumn 1994): 203–27.

Schaffer, Simon. "Paper and Brass: The Lucasian Professorship 1820–39." In Knox and Noakes, *From Newton to Hawking*, 241–93.

Schaffer, Simon. "Oriental Metrology and the Politics of Antiquity in Nineteenth-Century Survey Sciences." *Science in Context* 30, no. 2 (2017): 173–212.

Schweber, Silvian S., ed. *Aspects of the Life and Thought of John Herschel*. New York: Arno Press, 1981.

Schweber, Silvian S. "John Herschel and Charles Darwin: A Study in Parallel Lives." *Journal of the History of Biology* 22, no. 1 (Spring 1989): 1–71.

Secord, James A. *Victorian Sensation: The Extraordinary Publication, Reception, and Secret Authorship of Vestiges of the Natural History of Creation*. Chicago: University of Chicago Press, 2000.

Secord, James A. *Visions of Science: Books and Readers at the Dawn of the Victorian Age*. Oxford: Oxford University Press, 2014.

Sharp, Richard. "Wordsworth, Christopher (1774–1846)." *Oxford Dictionary of National Biography*.

Sheehan, Jonathan. *The Enlightenment Bible: Translation, Scholarship, Culture*. Princeton, NJ: Princeton University Press, 2005.

Shelford, April G. "Thinking Geometrically in Pierre-Damiel Huet's 'Demonstratio Evangelica' (1679)." *Journal of the History of Ideas* 63, no. 4 (2002): 599–617.

Siedentop, Larry. *Inventing the Individual: The Origins of Western Liberalism*. St. Ives, UK: Penguin Press, 2014.

Simpson, David. "Romanticism, Criticism and Theory." In Curran, *The Cambridge Companion to British Romanticism*, 1–24.

Sloan, Phillip R. "Whewell's Philosophy of Discovery and the Archetype of the Vertebrate Skeleton: The Role of German Philosophy of Science in Richard Owen's Biology." *Annals of Science* 60, no. 1 (January 2003): 39–61.

Slobodian, Quinn. *Globalists: The End of Empire and the Birth of Neoliberalism*. Cambridge, MA: Harvard University Press, 2018.

Smail, Richard. "Monk, James Henry (1784–1856)." *Oxford Dictionary of National Biography.*

Smith, Crosbie, and M. Norton Wise. *Energy and Empire: A Biographical Study of Lord Kelvin.* Cambridge: Cambridge University Press, 1989.

Smith, E. A. *A Queen on Trial: The Affair of Queen Caroline.* Stroud, UK: History Press, 2016.

Smith, F. B. "The Atheist Mission 1840–1900." In Robson and Cannon, *Ideas and Institutions*, 205–35.

Smith, Jonathan, and Christopher Stray, eds. *Teaching and Learning in Nineteenth-Century Cambridge.* Woodbridge, UK: Boydell & Brewer, 2001.

Smith, Jonathan, and Christopher Stray. *Cambridge in the 1830s: The Letters of Alexander Chisholm Gooden, 1831–1841.* Woodbridge, UK: Boydell & Brewer, 2003.

Smith, R. J. *The Gothic Bequest: Medieval Institutions in British Thought, 1688–1863.* Cambridge: Cambridge University Press, 1987.

Snyder, Laura J. *Reforming Philosophy: A Victorian Debate on Science and Society.* Chicago: University of Chicago Press, 2006.

Snyder, Laura J. "'The Whole Box of Tools': William Whewell and the Logic of Induction." In Gabbay and Woods, *Handbook of the History of Logic*, 163–223.

Snyder, Laura J. *The Philosophical Breakfast Club: Four Remarkable Friends Who Transformed Science and Changed the World.* New York: Broadway Books, 2010.

Stanley, Brian. "'Commerce and Christianity': Providence Theory, the Missionary Movement, and the Imperialism of Free Trade, 1842–1860." *Historical Journal* 26, no. 1 (March 1983): 71–94.

Stiglitz, Joseph E. *The Price of Inequality.* London: W. W. Norton, [2012] 2013.

Stone, Lawrence, ed. *Oxford and Cambridge from the Fourteenth to the Early-Nineteenth Century.* Princeton, NJ: Princeton University Press, 1975.

Stray, Christopher. "Curriculum and Style in the Collegiate University: Classics in Nineteenth-Century Oxbridge." *Histories of Universities* 16, no. 2 (2000): 183–218.

Stray, Christopher. "A Parochial Anomaly: The Classical Tripos 1822–1900." In Smith and Stray, *Teaching and Learning in Nineteenth-Century Cambridge*, 31–44.

Stray, Christopher. "One Museum to Another: The 'Museum Criticum' (1813–26) and the 'Philological Museum' (1831–33)." *Victorian Periodicals Review* 37, no. 3 (Fall 2004): 289–314.

Strong, E. W. "William Whewell and John Stuart Mill: Their Controversy about Scientific Knowledge." *Journal of the History of Ideas* 16, no. 2 (April 1955): 209–31.

Swade, Doron. *The Cogwheel Brain: Charles Babbage and the Quest to Build the First Computer.* London: Little, Brown, 2000.

Swade, Doron. "Calculation and Tabulation in the Nineteenth Century: Airy versus Babbage." PhD diss., University College London, 2003.

Thirlwall, John Connop, Jr. *Connop Thirlwall: Historian and Theologian*. London: Society for the Promoting Christian Knowledge, 1936.

Thompson, E. P. *The Making of the English Working Class*. London: Victor Gollancz, 1963.

Thompson, E. P. *Customs in Common: Studies in Popular Culture*. London: Merlin Press, 1991.

Trevelyan, G. M. *Trinity College: An Historical Sketch*. 1943. Reprint, Cambridge: Cambridge University Press, 1990.

Valone, David A. "Hugh James Rose's Anglican Critique of Cambridge Science, Antirationalism, and Coleridgean Idealism in Late Georgian England." *Albion: A Quarterly Journal Concerned with British Studies* 33, no. 2 (Summer 2001): 218–42.

Vance, Norman. "Farrar, Frederic William (1831–1903)." *Oxford Dictionary of National Biography*.

Verkaik, Robert. *Posh Boys: How the English Public Schools Run Britain*. London: Oneworld Publications, 2018.

Vidler, Alec R. *The Church in an Age of Revolution*. 1961. Reprint, London: Penguin Books, 1990.

Walsh, Harald T. "Whewell and Mill on Induction." *Philosophy of Science* 29, no. 3 (1962): 279–84.

Warwick, Andrew. *Masters of Theory: Cambridge and the Rise of Mathematical Physics*. Chicago: University of Chicago Press, 2003.

Wettersten, John. "William Whewell: Problems of Induction vs. Problems of Rationality." *British Journal for the Philosophy of Science* 45, no. 2 (June 1994): 716–42.

White, James F. *The Cambridge Movement: The Ecclesiologists and the Gothic Revival*. Cambridge: Cambridge University Press, 1962.

White, Paul. *Thomas Huxley: Making the "Man of Science."* Cambridge: Cambridge University Press, 2002.

Wilkes, M. V. "Herschel, Peacock, Babbage, and the Development of the Cambridge Curriculum." *Notes and Records of the Royal Society* 44, no. 2 (1990): 205–19.

Williams, Perry. "Passing on the Torch: Whewell's Philosophy and Principles of English Education." In Fisch and Schaffer, *William Whewell*, 119–46.

Willis, Kirk. "The Introduction and Critical Reception of Hegelian Thought in Britain 1830–1900." *Victorian Studies* 32, no. 1 (Autumn 1988): 85–111.

Willmoth, Frances, ed. *Flamsteed's Stars: New Perspectives on the Life and Work of the First Astronomer Royal, 1646–1719*. Woodbridge, UK: Boydell Press, 1997.

Wilson, Daniel C. S. "Arnold Toynbee and the Industrial Revolution: The Science of History, Political Economy and the Machine Past." *History & Memory* 26, no. 2 (Fall/Winter 2014): 133–61.

Wilson, David B. "Herschel and Whewell's Version of Newtonianism." *Journal of the History of Ideas* 35, no. 1 (January–March 1974): 79–97.

Wilson, David B. "William Whewell, Galileo, and Reconceptualizing the History of Science and Religion." *Notes and Records of the Royal Society* 65, no. 4 (December 2011): 343–58.

Wilson, Fred. "The Logic of John Stuart Mill." In Gabbay and Woods, *Handbook of Logic*, 229–81.

Withycombe, Robert S. M. "Perry, Charles (1807–1891)." *Oxford Dictionary of National Biography*.

Wolfe, Roger. "'Quite a Gem': An Account of the Former Mortuary Chapel at Mill Road Cemetery, Cambridge." *Cambridge Antiquarian Society* 84 (1996): 143–53.

Worthen, John. *The Cambridge Introduction to Samuel Taylor Coleridge*. Cambridge: Cambridge University Press, 2010.

Yanni, Carla. "On Nature and Nomenclature: William Whewell and the Production of Architectural Knowledge in Early Victorian Britain." *Architectural History* 40 (1997): 204–21.

Yeo, Richard. "Genius, Method, and Morality: Images of Newton in Britain, 1760–1860." *Science in Context* 2, no. 2 (Autumn 1988): 257–84.

Yeo, Richard. *Defining Science: William Whewell, Natural Knowledge and Public Debate in Early Victorian Britain*. Cambridge: Cambridge University Press, 1993.

Yeo, Richard. "Whewell, William (1794–1866)." *Oxford Dictionary of National Biography*.

Zarkadakis, George. *In Our Own Image: Saviour or Destroyer? The History and Future of Artificial Intelligence*. New York: Pegasus Books, 2015.

INDEX

Aarsleff, Hans, 79, 231
Affleck, Everena E., 107, 152
Affleck, Gilbert, 152
Agar, Jon, 138, 171, 232
Airy, George, 20, 85, 108, 133–35, 145, 225, 246
Albert, Prince (Francis Albert Augustus Charles Emmanuel), 106
algebra, 8, 10, 13, 20–25, 27–29, 31, 34, 36, 38, 44–46, 53–54, 57, 65, 67, 102, 106, 119, 131, 133, 138, 144–45, 158, 233, 236, 240, 242
Allen, John, 147
Analytical Engine, 28, 33, 132, 155
Analytical Society, 21–22, 28–29, 53, 119, 134–35, 232, 236
Anti-Corn Law Movement, 11, 96
Apostolic, 67, 86
a priori, 7, 13, 44, 95, 115, 117–18, 122
Arbuthnot, John, 114
Aquinas, Thomas, 110
Arianism, 15
Arnold, Thomas, 42–43, 73, 75, 103, 131
artificial intelligence, 29, 56, 172, 234, 236, 247
Ashley, William J., 100
Astronomical Society of London, 13, 23, 57, 73, 111–12, 114, 134
astronomy, 13, 24–25, 45, 110–11, 135, 144, 230, 232, 238
atheism, 24, 112, 135, 162
Athenaeum, The (Cambridge), 39, 46, 230
Athenaeum, The (German), 36
Babbage, Charles, 20–21, 23–28, 32–33, 36, 44–46, 53–54, 56–57, 59, 68, 102, 109, 114–15, 117–19, 131–35, 137, 143–44, 155, 159, 162, 171, 225–26, 230, 232, 236–37, 240, 244–46
Bacon, Francis, 49, 115–16, 136
Baily, Francis, 111–14, 124, 226, 232
Barrow, Isaac, 116, 136, 231
Barrow, John, 73, 114
Barnwell, 26, 103
Bayne, William Thirlwall, 49
Becher, Harvey, 24, 232–33
Benthamism, 8, 41, 105, 121–22, 126, 171
Bentham, Jeremy, 16, 42, 56, 60, 67, 79, 123, 137–38, 143, 168, 226
Bentley, Richard, 76
Berkeley, George, 29
biology, 13, 242, 244
Berlin University, 64, 66
Beveridge, William, 165
Bishop Maltby (Edward Maltby), 79
Blakesley, Joseph Williams, 49
Blomfield, Charles James, 36, 76, 78–79, 101, 160
Boeckh, August, 70
Boisseree, Sulpiz, 141
Boole, George, 28, 226, 239
Bonaparte, Lucien, 54
Bopp, Franz, 77
Boyce, Edward Jacob, 46
Bretschne, Karl Gottlieb, 69
Brewster, David, 113

Bridgewater Treatise, 28, 32, 45, 101, 110, 116, 225, 230
Bright, John, 96
British and Foreign Bible Society, 10, 52–53, 64, 228
British Critic, 37, 89
Bromhead, Edward, 23, 28, 97
Brougham, Henry, 13, 132
Buller, Charles, 91
Bullock, Charles, 162
Bulwer-Lytton, Edward, 49
Bunsen, Karl Josias von, 38–40, 44, 58, 65, 70–71, 75, 77, 83–86, 91
Burdett, Francis, 14–15
Burke, Edmund, 74, 121
Burkean, 165, 170
Burrow, John Wyon, 77, 156, 233
Butler, Joseph, 121, 123, 231

calculation, 4, 16, 55–56, 59, 79, 128, 130, 132, 136, 169, 171, 246
Cambridge Apostles, 39, 41–42, 49–50, 78, 232, 235
Cambridge Camden Society, 46–48, 122, 141, 143
Cambridge Etymology Society, 76, 79
Cambridge riots, 37
Cambridge Union, 19, 21
Carlyle, Richard, 81–82
Carlyle, Thomas, 146–47, 165, 234, 239
catastrophism, 80, 155
Cathedral Commission, 146
Cato Street Conspiracy, 30
Cayley, Arthur, 28
Chadwick, Edwin, 105, 238
Chadwick, Henry, 30, 234
Chadwick, Owen, 47, 89, 234
Chambers, Robert, 72, 242
Charterhouse School, 9, 19, 36, 49, 62, 73, 165, 229
Chartism, 11, 28, 100, 147–48
chemistry, 13, 58, 107
Christian Remembrancer, 37
Christian Socialist Movement, 147–48
Christ's Hospital School, 68
Church of England, 4, 7–8, 10–19, 24, 28, 35–37, 50–53, 67–68, 70, 76, 78, 80, 82, 89, 93, 95–96, 99, 118, 146, 149–50, 153, 168, 237
Church Missionary Society (1799), 35, 148
Clarke, Edward Daniel, 9, 52
Clarke, Samuel, 16
Classical Tripos, 38, 42, 76, 78, 87, 245
coaching, 108, 124, 132, 134–35, 139, 167
Cobden, Richard, 96, 100
Coddington, Henry, 87
Colchester Royal School, 20
Coleridge, Derwent, 49
Coleridge, Samuel Taylor, 14–16, 24, 39–43, 47–48, 50–51, 56, 58–60, 75, 78, 80, 109, 117–18, 123, 131, 157, 226, 238, 240, 242–43, 246–47
Coleridgean, 42, 138
Comte, Auguste, 117, 139, 231
Comtean, 138
Condillac, Etienne Bonnot de, 23, 25, 226
Conybeare, William Daniel, 99
creationism, 3, 236
Cunningham, William, 100, 168–69, 240

Darwin, Charles, 33, 81, 122, 153–56, 158, 160–62, 232, 235, 244
Darwin, Erasmus, 233
deism, 15, 53
Denison, William, 162
Desmond, Adrian, 13, 153–54, 235
Dickens, Charles, 62, 92
Difference Engine, 28, 132
Digby, Kenelm Henry, 19, 48, 241
Dissenters (religious), 10–11, 13–14, 17, 35, 52, 76, 81–83, 85–91, 96, 99, 109, 149, 151, 162, 230
divine design, 3–4, 13, 16, 27, 50, 53, 55, 115, 170, 235–37
Dobree, Peter Paul, 68
Donaldson, John William, 78, 237
Donne, William Bodham, 49–50, 78–79
D'Oyly, George, 37
Dragon School, 168
Drinkwater, John Elliot, 100

Duke of Wellington (Arthur Wellesley), 61
Durham University, 136, 163
Dymoke, Lionel, 146

Earl of Powis (Edward Herbert), 106
East India College, 35, 61, 123, 139
Ecclesiastical Revenue Commission (1835), 83
Edinburgh Review, 98, 228
Edinburgh University, 31, 135, 168, 226, 241
Eichhorn, Johann Gottfried, 65
Elliott, Edward Bishop, 19
Ellis, Francis, 152
Ellis, Robert Leslie, 49, 152
Ely, 30
Ely Cathedral, 143
Empson, William, 123
English Civil War, 14
epistemology, 3, 12, 15, 41, 113–14, 154–55
ethical economy, 17, 60, 70, 77, 94, 96–97, 120, 165–66
Eton School, 15, 19–20, 31, 35, 37, 49, 74, 76, 79, 120, 145, 151, 166, 168
Etymology, 40, 75, 77, 80, 239
Euclid, 22, 34, 108, 119, 124, 135, 151
Euclidian, 8, 22, 29, 34, 108, 110, 140, 152, 158, 167
evolution, 33, 39, 67, 72, 79, 153–56, 160, 165, 170–71, 223, 235, 237

Fall (of man), 3, 26, 29, 53–54, 60, 69, 111, 126, 154, 170, 238
Farish, William, 52
Farrar, William Frederic, 78, 226, 246
Fichte, Johann Gottlieb, 40, 233
Fitzgerald, Edward, 49–50, 78
Fitzwilliam Museum, 142
Flamstead, John, 111–14, 121, 226, 230, 232, 246
Forbes, Duncan, 41
Forbes, James David, 133, 135, 142, 152, 226
French Revolution, 8, 10, 13, 15, 26, 30, 36, 44, 74–75, 109, 121, 170, 235, 237
French, William, 21, 122
Frend, William, 15, 48, 51
Froude, Hurrell, 89, 151, 237
fundamental ideas, 7, 22, 31–32, 34, 43, 46, 57–58, 62, 70, 72, 80, 94, 112, 114–15, 117–18, 120, 127, 131, 140, 143–44, 169

Galilei, Galileo, 110, 247
Garnham, Robert Edward, 50, 225
genius, 8, 17, 20, 39, 45–46, 60, 109, 111–14, 137, 171, 232, 244, 247
Geological Society of London, 13, 153
geometry, 8, 22, 24, 27–29, 31–32, 34–35, 71, 106, 108, 110–11, 115, 124, 131, 133, 139–40, 152, 158, 167, 231, 242
German Idealism, 43, 233
Gladstone, William, 145, 226
Glasgow University, 23, 39, 133
Goethe, Johann Wolfgang von, 42, 66, 242
Gooden, James, 19, 245
Goodwin, Charles Wycliffe, 161
Gossett, Thomas Stephen, 102
Gothic, 19, 35, 46–47, 100, 122, 139–43, 153–54, 245–46
gradualism, 155
Great Court, 84, 142, 146
Great Gate, 142
Great Reform Act (1832), 11, 14, 61, 83, 136
Green, Thomas Hill, 164, 166–67, 234, 243
Gregory, David, 114
Gregory, Duncan Farquharson, 28
Grimm, Wilhelm Carl, 77
Grote, George, 73

Hallam, Arthur Henry, 49
Hallam, Henry, 31, 122
Halley, Edmund, 112–14, 228
Hamilton, William, 31, 226, 233, 239
Hardenberg, Friedrich von (Novalis), 40

Hare, Augustus John Cuthbert, 48
Hare, Augustus William, 41
Hare-Naylor, Francis, 9
Harrison, Frederic, 161
Harrison, Peter, 110, 168, 238
Harrow School, 42, 46, 48, 88, 91, 147
Hartley, David, 34
Hawkins, Edward, 32, 105, 145
Hawtrey, Edward Craven, 79, 149
Hayek, Friedrich August von, 169–71, 233–34, 238–39
Heath, Douglas Denon, 49
Hegel, Georg Wilhelm Friedrich, 44, 231, 246
Hengstenberg, Ernst Wilhelm, 66
Henslow, John Stevens, 155
Herder, Johann Gottfried, 44, 66
hermeneutics, 65, 67, 238
Herschel, John, 20–21, 23–28, 33, 43, 53–54, 57, 68, 107, 114–15, 117–20, 134, 137, 155, 225, 227, 232, 234, 244, 246–47
Higgit, Rebekah, 113, 238
Hill, Christopher, 166
Hilton, Boyd, 185, 238
Hobbes, Thomas, 41–42
Hole, Robert, 16, 238
Holland, Henry Scott, 166
Hooker, Joseph Dalton, 153
Hope, Alexander Beresford, 46
Hopkins, William, 133–34
human logic, 8, 10, 24, 27, 29–31, 57, 67, 143, 164, 167, 170
Humboldt, Alexander von, 77
Hume, David, 8, 32, 112, 137, 170, 227
Hunt, Henry, 21, 30, 235
Huxley, Thomas Henry, 154–55, 235, 246
Hyde Hall, 61

imagination, 5, 8, 20–21, 39–41, 43, 56, 58–60, 71, 75, 115–16, 137, 153, 232, 238, 240
incarnation, 164, 167
innate ideas, 7–8, 17, 22, 27, 50, 105, 112, 115, 117–19, 137, 153, 157, 170
Institution for Promoting the British System for the Education of the Labouring and Manufacturing Classes of Every Religious Persuasion (1814), 53
intellectual property, 114
International Law, 69, 107, 168, 170
intuition, 39, 43, 70, 81, 110, 117, 118

Jacob, Edward, 22
Jacob, William, 22
Jacobi, Friedrich Heinrich, 44, 233
Jacobinism, 31, 52, 74
Jephson, Thomas, 25
Jeremie, James Amiraux, 139
Jones, Richard, 17, 23, 32, 41, 43–44, 60, 70, 73, 77, 79–80, 83–84, 88, 91, 93–100, 103, 112, 117, 119–20, 122, 128, 134, 141, 144–47, 152, 169, 225, 227–28, 237, 242
Jones, Thomas, 52
Jones, William, 77, 236
Jowett, Benjamin, 160, 163–64

Kant, Immanuel, 7–8, 40, 42–44, 58, 77, 102, 107, 109–10, 112, 115, 170, 232, 235, 239–40
Kantian, 7–8, 44, 115, 153, 157, 170
Kaye, Charles, 38
Kaye, John, 76
Kemble, John Mitchell, 42, 50, 78
Kepler, Johannes, 110–12, 116–17
King Edward VI Schools, 20 (Totnes), 36, 37 (Cheltenham), 42 (Louth), 49–51, 78 (Bury St. Edmunds), 79 (Birmingham)
King's College London, 37–38
King's School, 15, 51
Kingswood School, 168

Lacroix, Sylvestre, 21–23, 53, 139
Lagrange, Joseph-Louis, 21, 23–24, 28, 32, 139
Lamark, Jean-Baptiste, 13, 162
Lancaster, Joseph, 37, 52
Laplace, Pierre-Simon, 21, 23, 45, 139

Latham, Robert Gordon, 79, 243
Lawrence, William, 56
Lax, William, 25
Leask, Nigel, 65
Lee, James Prince, 79
Leibniz, Gottfried Wilhelm, 16, 28
Leibnizian, 21
Lewis, George Cornewall, 74
liberal education, 33, 58, 86, 105, 108, 111, 115, 123, 129, 131, 134–35, 138, 152, 230–31
liberalism, 5, 94, 244
Liebig, Justus von, 117
Lincoln Mechanics Institute, 28
Linker, Damon, 44, 240
Lipson, Ephrain, 100
Littleport riots, 30
Locke, John, 8, 15, 20, 23, 34, 42, 71, 112, 136–38
Lockean, 15, 72, 123, 137
London School of Economics, 169
London Society for Promoting Christianity amongst the Jews (1809), 35
London University, 13–14, 50, 86, 91
Long, George, 19
Lonsdale, John, 37
Lord Kelvin (William Thompson), 133, 245
Lord Liverpool (Robert Jenkinson), 30, 37, 52
Lucasian Professor of Mathematics, 25, 45, 52, 85, 113, 131–32, 239, 244
Lucke, Gottfried Christian Friedrich, 65
Ludlow, John Malcolm, 147–48
Luther, Martin, 110
Lutheran, 7, 75
Lyell, Charles, 106, 155–56, 158, 227, 243
Lyttelton, Henry William, 150

Macaulay, Thomas Bebington, 49, 116, 231
Macintosh, James, 121–23, 227
Magee, William, 143
Maitland, William, 168, 240
Malcolm, Isabella (Lady Malcolm), 61
Malcolm, John, 61
Malcolm, Kate, 61
Malden, Henry, 49
Malkin, Benjamin Heath, 50, 78, 241
Malthus, Thomas, 44, 99
Mandeville, Bernard, 55, 119, 170
Manners-Sutton, Charles, 37
Manning, Henry Edward, 147
Mansel, William Lort, 37
Marsh, Herbert, 51–52, 64, 228–29, 236
Marshall, Alfred, 105, 169
Marshall, Cordelia, 72
Marshall, James, 12, 80
Marshall, John, 72
Marshall, Susan Harriet, 72
Marx, Karl, 97, 144, 168, 228, 237
Mathematical Tripos, 20, 24, 38, 131, 237
materialism, 8, 15–16, 39, 56, 70, 79, 110, 152, 236
Maule, William Henry, 135
Maurice, Frederick Denison, 39–42, 78, 123, 147–48, 157, 161, 169, 228, 233, 235, 242–43
McCulloch, John Ramsay, 94, 98, 228
McFarland, G. F., 70, 240
Mead, Richard, 114
Melanchthon, Philipp, 110
Melbourne, Second Viscount (William Lamb), 91
Merchant Taylors School, 32, 105
Merivale, Charles, 42, 99
metaphysical, 3, 5, 7, 43–44, 115, 118, 165, 169
metrology, 77, 244
Michaelis, Johann David, 64
Mill, James S., 56, 67, 122
Mill, John S., 4, 40–41, 67, 71, 79, 91, 94,111, 116–18,124, 137, 143, 146, 228, 242, 245–47
Mill Road Cemetery, 142, 247
Mill, William Hodge, 47, 79
Miller, William Hallowes, 61
Milner, Isaac, 25–26, 52, 228, 239
Moller, Georg, 141
Monk, James Henry, 36, 38, 40, 48–49, 76, 245

Moore, James, 154, 235
Moral Tripos, 106–7
Morgan, Augustus De, 28, 112, 115, 241
Morris, William, 165
Moultrie, John, 49
Muller, Karl Otfried, 70
Muller, Max, 77
Musgrave, Thomas, 88
Myers, Arthur Thomas, 72
Myers, Frederic William Henry, 72, 124

National Society for the Education of the Poor in the Principles of the Established Church (1811), 37, 53
Natural Sciences Tripos, 106–7
natural selection, 3, 33, 67, 156, 163
Neale, John Mason, 46
Neanderthal, 156
neoliberalism, 169, 171, 233, 235, 239, 244
New Court, 38, 139, 142
Newman, John Henry, 30, 47, 89, 105, 123
Newton, Isaac, 17, 22–24, 28, 35, 45, 85, 106, 110–14, 116, 133, 135–36, 145, 230–31, 237–39, 244
Newtonianism, 21–22, 24, 34, 57, 109, 113, 136, 238, 247
Niebuhr, Barthold Georg, 39, 43, 65, 70–71, 73–76, 123, 227, 233
noblemen, 9
Norris, Henry Handley, 37

O'Connell, Daniel, 99
O'Connor, Fergus, 100
Okes, William, 21
ontology, 43, 72, 115
Oriel Noetics, 12, 86
Owen, Richard, 153, 244
Oxford Movement, 40, 47–48, 67, 105, 226, 236

Paine, Thomas, 53, 228
Paley, William, 4, 15–16, 32, 40, 42, 48, 55, 109, 119, 123, 125–27, 136–39, 228, 233, 236–37
pantheism, 65
Parker, John William, 148, 160, 228
Patteson, John, 37
Pattison, Mark, 160, 236
Paulus, Heinrich Eberhard Gottlob, 65
Peacock, David Mitford, 33, 228
Peacock, George, 20, 23, 25, 88, 102, 106, 134, 143, 169, 233, 240, 242, 246
Peel, Robert, 101, 103–4, 135, 243
pensioner, 9
Perry, Robert, 88, 92, 200, 247
Peterloo Massacre, 30, 147
Philological Museum, 74, 76–77, 85, 235, 245
Philological Society of London, 79
philology, 39, 60, 75–78, 80, 85, 108, 156, 233
Philpott, Henry, 103
Pinkard, Terry, 44, 64, 241
Plato, 15, 40, 77, 110, 159, 231
plurality of worlds, 129, 231, 235
poetry, 15, 20, 39–44, 59–60, 78, 105, 108, 123, 131, 235, 242
pointed arch, 139–41, 153
Polanyi, Karl, 168–69, 243
political economy, 8, 12, 17, 32, 43–44, 55, 58, 60, 71, 77, 79, 85, 93–99, 119–20, 143–44, 170, 174, 225, 227–28, 230–31, 239–41, 247
Poor Law Amendment Act (1834), 143
Porson, Richard, 75
Potato Famine, 103
Powell, Baden, 117, 160, 162, 234
Praed, Winthrop Mackworth, 49
Preyer, Robert O., 80, 242
Pritchard, James Cowles, 77–78
proctor, 19, 61, 81
providence, 3, 16, 40–41, 51, 54, 69, 73, 111, 136, 149, 167, 245
Prussian Correspondent, 74
Pryme, George, 49, 85, 88, 95, 240
Pusey, Edward, 47, 123

Quaker, 37, 74, 146
Queen's College, Cork, 28

railway, 147, 152
Raine, Matthew, 9
Ray, John, 136
Rebecca Riots, 147
redemption, 7, 67, 126, 159
revelation, 39, 46, 67
Ricardian, 8, 56, 77, 79, 94–8
Ricardo, David, 60, 94–96, 98, 118–19, 143, 167–69, 230
Rice, Thomas Spring, 85
Richards, Joan, 22, 117, 242
Richards, Robert J., 154, 242
Richter, Melvin, 165, 243
Rickman, Thomas, 140–41
Rigaud, Stephen Peter, 113, 228
Robinson, John, 31
Rogan, Tim, 165, 243
Rousseau, Jean-Jacque, 50
Royal Commission Inquiring into the Cambridge and Oxford Curriculum, 106, 134
Royal Cornwall Polytechnic Society, 146
Royal Dragoons, 30
Royal Society of London, 13, 44
Rugby School, 75, 160, 165
Russell, John, 99, 103

salvation, 7, 36, 54–55, 113, 125, 129, 154, 159, 164–66
Salvin, Anthony, 141–42
Saunders, George, 140
Scarlett, James, 135
Schaffer, Simon, 104, 217, 244
Schelling, Friedrich, 40, 44, 66, 115, 153, 233, 240
Schlegel, August, 36
Schlegel, Friedrich, 36, 77
Schleiermacher, Friedrich, 40, 64–65, 86, 162, 229
Scholefield, James, 68
Scott, George Gilbert, 143
Scott, Walter, 48
Scripture, 4, 10, 17, 24–25, 50–51, 54, 60, 62–3, 65, 67–9, 82, 122, 127, 129, 146, 148, 159, 162–63, 165–66
Sedbergh School, 9, 20, 49
Seditious Meetings Act (1817), 22
Sedgwick, Henry, 168
self-interest, 14, 16, 119, 167, 169–70
selfish, 11, 16, 19, 41–2, 47, 55, 62, 84, 148, 235
Senate House, 61, 124
Senior, Nassau William, 94, 98, 120
Shakespeare, William, 42, 60
Sheepshanks, Richard, 20, 30, 88, 134, 238
Sherbourne School, 46
Shipley, Georgina, 9
Simeon, Charles, 35, 52, 61, 148, 238
Slobodian, Quinn, 170, 244
Smith, Jonathan, 190, 245
Smith, Sidney, 103
socialism, 147, 165, 234
socialist, 147–48, 165–66, 168
Society for the Diffusion of Knowledge, 74
Society in Aid of the Incorporated Society for the Propagation of Gospels in Foreign Parts (1842), 122
Socinianism, 15
soul, 7, 10, 12, 15–16, 20, 24, 34, 43, 50, 56, 58–60, 69, 80, 83, 111–13, 115–16, 119, 126, 128–29, 133, 136–37, 148, 154, 157–58, 167, 170
South, James, 134
Spedding, James, 49–50, 78
Spence, Thomas, 21
statutes, 102–3, 131, 145
Sterling, John, 39–40, 42, 146–47, 229, 234
St. John's School, 149
St. Paul's School, 49, 79
Stray, Christopher, 38, 76, 176, 210, 245
Stuart, Margaret Brodie (Margaret Herschel), 54
sub-sizar, 9
Swing Riots, 13, 83–84
Sylvester, James J., 151, 229

Tambroni, Clotilde, 9
Tawney, Richard H., 165–69, 239, 243

Taylor, Robert, 81
Taylor series, 24
Temple, Frederick, 160
Temple, William, 165
Ten Hour Factory Act (1847), 147
Tennyson, Alfred, 42, 116
Test and Corporation Act (1661 and 1673), 14, 82
Tindal, Matthew, 15
Tindal, Nicholas Conyngham, 37
tithe, 13, 83–84, 96, 99–100, 147, 227, 229
Tithe Commutation Act (1836), 99–100
Thompson, Edward P., 166, 168–69, 241, 243, 246
Thompson, William Hepworth, 163
Thorp, Thomas, 19, 47, 237
Toland, John, 15
Tooke, Horne, 15, 78–79
Townsend, Charles, 97
Toynbee, Arnold, 166–67, 169, 229–30
Tractarians, 12, 30, 47–48, 65, 89, 91, 102, 106, 144, 149, 151
Trench, Richard Chenevix, 42, 105, 123
Trevelyan, George Macaulay, 131, 246
Trinity Chapel, 125, 128, 157
Trinity Lodge, 38, 141–42
Troughton, Edward, 134–35
Turton, Thomas, 85

Uckfield School, 9
uniformitarian, 155
uniformitarianism, 80, 155
Unitarian, 15, 28, 36, 48, 52, 65, 106, 111–12
Unitarianism, 15, 48, 112
University College, London, 14, 19, 28, 48–9, 78
utilitarian, 4, 11, 15, 17, 27, 29, 33–4, 39, 41–2, 44, 47, 53, 56, 60, 73, 79, 82–3, 97, 101, 105, 107, 123, 128, 130–32, 136–38, 144, 149, 152, 159, 167–68, 171
utilitarianism, 4, 6, 8, 41, 48, 55, 65, 70, 74, 80, 116, 125–26, 132, 165, 167, 169, 171

Vansittart, Nicholas, 34
verticality, 57, 115, 140–41, 153
Vince, Samuel, 25

Waddington, George, 19
Waddington, Horace, 19
Wade, John, 82
Walker, William Sidney, 19
Warwick, Andrew, 133, 246
Watson, Joshua, 53
Watson, Richard, 52
Wells, William Charles, 117
Westminster School, 22, 99–100
Wette, Wilhelm Martin Leberecht de, 65
Whately, Richard, 118–20, 143–44, 170, 230, 236–37
Whewell, Ann, 58
Whewell, Cordelia, 72, 143, 145, 152
Whiston, William, 114
Wilberforce, Robert, 144
Wilberforce, Samuel, 147, 161–62
Wilberforce, William, 121, 144
Wilkins, William, 139
Williams, Rowland, 161
Willis, Robert, 140, 233
Wilson, Henry Bristow, 160
Winchester School, 36, 85, 149, 155
Wolf, Friedrich August, 73
Wood, James, 21–22, 25
Woodard, Nathaniel, 149
Wordsworth, Cannon Christopher, 85, 161
Wordsworth, Christopher, 36–38, 48, 52–53, 66, 76, 87–88, 92, 101, 122, 139, 142, 231
Wordsworth, William, 9, 36, 39–42, 47, 51–52, 55, 57–60, 123, 139, 242–44
Wordsworthian, 20, 36, 39, 42
Worsley, Thomas, 122–24
Wren, Christopher, 114

Yeo, Richard, 33, 247
Young, Thomas, 31